Smart Coatings

This book focuses on fundamentals, technology, synthesis, and characterizations and applied techniques from a practical point of view of coatings. The first three chapters offer a rigorous review of the application of these coatings to corrosion protection in various aerospace and oil and gas industries, and the subsequent chapters present a quick critical review of the state-of-the-art protection techniques of these coatings to novel biomedical applications such as clinical translations and tissue-engineered materials. Environmental, ergonomics, and aesthetic aspects and future perspectives are also explained at the end.

Features:

- Explores the synthesis and application techniques of novel smart coatings in various research areas
- Presents a concise, critical, and state-of-the-art review of existing research on various types of smart coatings
- Ascertains the different mechanisms associated with the stimuli response of smart coatings
- Includes an exclusive chapter on real-time applications in the biomedical field
- Covers self-healing, self-cleaning, pH balance, early corrosion detection, and triggering mechanisms

This book is aimed at researchers and graduate students specifically in smart coatings and thin films and corrosion, including chemical, materials science engineering, industrial and manufacturing engineering, and nanotechnology.

Smart Coatings
Fundamentals, Developments, and Applications

Vaibhav Sanjay Kathavate
Pravin Pralhad Deshpande

CRC Press
Taylor & Francis Group
Boca Raton London New York

CRC Press is an imprint of the
Taylor & Francis Group, an **informa** business

First edition published 2023
by CRC Press
6000 Broken Sound Parkway NW, Suite 300, Boca Raton, FL 33487-2742

and by CRC Press
4 Park Square, Milton Park, Abingdon, Oxon, OX14 4RN

CRC Press is an imprint of Taylor & Francis Group, LLC

© 2023 Vaibhav Sanjay Kathavate and Pravin Pralhad Deshpande

ISBN: 978-1-032-06079-8 (hbk)
ISBN: 978-1-032-06080-4 (pbk)
ISBN: 978-1-003-20063-5 (ebk)

DOI: 10.1201/9781003200635

Typeset in Times
by codeMantra

Contents

Preface

"What is the greatest gift to human society in the 21st century?" Of course, the "Materials". The word "Materials" serves diverse choices of constituents ranging from engineering materials such as wood (naturally occurring composite) and conventional steel to functional nanomaterials (or thin films) in micro-electro-mechanical systems (MEMS) and nano-electro-mechanical systems (NEMS) devices. However, one common question to pose at this juncture is: With ever increase in miniaturization and automation, do these materials have the potential to solve the great and pressing problems of society? And if the answer to the above question is "Yes", is it possible to make them auto-responsive when the need arises? The surface functionalization of materials *via* "Smart Coatings" is the answer to the latter one. These smart coatings auto-respond to the changes occurring around the surrounding environment (in both automatic and external triggering modes) and prevent external damage to the underlying material. For instance, stimuli-responsive smart coatings can detect the corrosion events occurring in metals at their early stages and take the preventing measure. Owing to their excellent self-healing, self-cleaning, and sensing characteristics, these smart coatings find a unique niche in widespread surface functionalization applications, thus a topic of systematic and considerable research.

Although these smart coatings have been researched for the past 3–4 decades, their practical applications on the commercial scales are still far from industrial acceptance. When Dr. Deshpande put forth this idea of writing a book on "Smart Coatings", the immediate thought that came to my mind was the applicability and advances in these coatings for the diverse range of applications. Keeping this in view, we are pleased to present the book *Smart Coatings: Fundamentals, Developments, and Applications* to the materials science and coatings community. While reviewing the articles for this book, we have noticed that more emphasis was given on the synthesis and applying techniques previously, and advancement in the field with the current issues is seldom reported. In view of this, the present book provides significant insights about the current issues, various advancements and diverse range of applications in particular, and synthesis and applying (deposition) techniques in general.

The organization of chapters and their subsequent contents is made so that it will provide a comprehensive review of the applications of smart coatings in various sectors starting from fundamentals to recent advancements in the field. For instance, each chapter of the present book begins with a general introduction followed by a strong motivation that helps the scientist, researchers (research scholars also), and academician to understand the importance and depth of the subject from various applications' viewpoints, while the current issues/challenges and future perspectives based on the authors' own views are outlined in the last part of the chapters. The purpose of including challenges and future scope is to guide the research community about the current hurdles faced during the applications of these coatings in various sectors such as automobile, marine, defence, biomedical (i.e., medical implants), and clinical translations. This also pinpoints some important open questions in the field that can be addressed in future and open up new pathways to tailor the multifunctionality in these coatings. Towards this end, the authors have profoundly reviewed the articles till June 2021.

The first chapter provides a general overview of the coatings, their historical developments, and recent advancements from various applications' viewpoints. The current issues in the present coatings technologies are outlined, thus paving a motivation for stimuli-responsive smart coatings. Different stimuli-response mechanisms are highlighted in the middle, followed by recent advancements. Following this, the second chapter presents a brief overview of the electrochemical aspects of corrosion in metals (and alloys) in general and its prevention strategies *via* surface functionalization in particular. A complex mathematical treatment related to the kinetics of corrosion is avoided. Rather more emphasis is given to corrosion prevention by smart coatings. Chapter 3 of this book addresses the applicability of these coatings in the automobile, aerospace, and defence sectors. This chapter starts with the advancements in conventional materials and respective surface preparation strategies. Water-soluble and UV-curable smart coatings with their multifunctional characteristics and related triggering mechanisms are highlighted in the middle. Finally, optimization of process parameters by synthesis and applying techniques of these coatings for the mentioned applications is outlined. Subsequently, Chapter 4 connects the real-time applications in biomedical fields to various triggering mechanisms offered by these coatings. The shape memory alloys/polymers and nanobioceramic smart coatings for bio-implant applications are briefly outlined on this front. Following this, Chapter 5 provides significant insights into challenges related to applying these coatings in clinical translations. Towards this end, the release mechanisms such as active triggering and triggered release are outlined with current challenges and future perspectives. Chapter 6 highlights the synthesis, applying techniques, and triggering mechanisms related to smart switchable coatings from a marine and medical applications viewpoint. The surface functionalization *via* switchable smart coatings is of paramount importance in water/oil repellent applications in marine industries and controlled drug delivery at therapeutic sites in medical sectors. Such unique aspects of smart coatings are explored in Chapter 6. Finally, Chapter 7 gives closure and sheds light on the environmental aspects, followed by a general discussion on the challenges related to the enhancement in multifunctionality of smart coatings. All the chapters are organized concisely, and the attempts are made to provide succinct information about every aspect.

We believe that the present book would be helpful to the scientists, researchers (research scholars), and academicians from diverse research backgrounds such as biomedical science, civil and structural engineering, mining and marine engineering, metallurgy engineering and materials science, mechanical engineering, and nanoscience and nanotechnology. Note that the explanations and expressions in this book are the author's views based on expertise in the field. However, readers are also encouraged to have a detailed look (if they want) at the cited references to understand the subject further. We have also attempted an interdisciplinary approach to streamline the present work for bridging the gap between various science and engineering disciplines.

Of course, to streamline this work, many helping hands led to the timely completion of manuscripts, artwork, and type settings. Acknowledging them is only a small thing we can do in return for their help. First and foremost, both the authors are thankful to the editorial crew of CRC Press for accepting the proposal and availing the gracious opportunity to write this book. Dr. Pravin P. Deshpande would like to thank Prof. B. B. Ahuja, Director of Engineering Pune, for his constant support. He also would like to thank his colleagues and associates from the Department of Metallurgy and Materials Science, College of Engineering Pune. Dr. Vaibhav S. Kathavate is indebted to the Director, Indian Institute of Technology Indore (IIT Indore) and Indian Institute of Technology Bombay (IIT Bombay) for their continuous encouragement. The moral support from Dr. Eswara Prasad Korimilli, IIT Indore, Cdr. K. Gopkumar, NIOT Chennai, and Prof. A.S. Adkine, Hitech Institute of Technology, Aurangabad is also appreciated. Thanks are also due to Dr. Niteen G. Jadhav, Nelumbo, Inc., California, USA for the helpful discussion and sincere suggestions during the proposal stage. We are grateful to various publishing houses and copyright clearance centre (CCC) and RightsLink® for helping to provide us with the copyright permissions for reproducing and reprinting the previously published contents in this book.

Lastly, both the authors would like to thank their families and friends for the continuous encouragement and patient support during the course of work. VSK is also grateful to Dr. Pravin P. Deshpande for the helpful discussion and his timely advice during the writing stage of manuscripts.

To sum up, it gives us immense pleasure to present our book *Smart Coatings: Fundamentals, Developments, and Applications* to the readers. We hope that the readers from diverse backgrounds can enjoy the reading and learn something meaningful at the end of the chapters. Lastly, the present book in single *Layman's term* can be described as "21st Century Smart Coatings: Where We Are? And a Way Forward".

Vaibhav S. Kathavate

Authors

Vaibhav Sanjay Kathavate, PhD, landed at the North Caroline State University (NCSU), USA in May 2022 after his first postdoctoral research at the Indian Institute of Technology Bombay (IITB). He earned his PhD from the Indian Institute of Technology Indore (IIT Indore). At NCSU, he is associated with the Department of Mechanical and Aerospace Engineering as a Post-Doctoral Research Fellow. Prior to joining IIT Indore, he served as Project Scientist at the National Institute of Ocean Technology Chennai (NIOT) under the aegis of Ministry of Earth Sciences, Government of India. His research interests lie at the intersection of experimental mechanics and micro/nanotechnology, electrochemistry at surfaces and interfaces of advanced crystalline materials, and stimuli-responsive smart coatings for clinical translation and water/oil repellent applications. With a continuing motivation from "nature", he is keen to correlate his research with natural activities, thereby serving the emerging needs of the society. He has authored various scientific articles of an international repute. When Dr. Kathavate is not researching, he attempts to sing classical songs involving different Indian languages. He is also enthusiastic about experiencing different cultural and regional activities across the nooks and corners of the globe.

Pravin Pralhad Deshpande, PhD, is an Associate Professor in the Department of Metallurgy and Materials Science, College of Engineering Pune (COEP), India. He has completed a number of research projects sponsored by the All India Council for Technical Education (AICTE), University Grants Commission (UGC), Indian Space and Research Organization (ISRO), Indian National Science Academy (INSA), and University of Pune (UoP). He has authored many referred international publications, guided PhD students from diverse background, and worked as a consultant for many industries. He is an active member of a study group on

Metallurgical Heritage of India, Indian National Academy of Engineering, India. He served as a Visiting Professor at Peter the Great Saint Petersburg Polytechnic University (SPbPU), Russia during 2016 and 2017. His hobbies include playing electrical benjo and reading literature on ancient metallurgy.

Abbreviations and Symbols (Units)

1-D	One dimensional
2-D	Two dimensional
3-D	Three dimensional
C	Coulomb
cm	Centimetre (cm)
E_0	Standard electrode potential (V)
$E_{(0(a))}$	Standard electrode potential at the anode (V)
$E_{(0(c))}$	Standard electrode potential at the cathode (V)
E_{corr}	Corrosion potential (V)
e^-	Free electrons
E_m	Potential of the corroding substrate/metal (V)
EPR	Enhanced permeability and retention
F	Faraday's constant (C/mol)
ΔG	(Gibbs) free energy
GPa	Giga Pascal
HPO_4^{2-}	Phosphate
i_{corr}	Corrosion current density (A/cm^2)
i_e	Exchange current density (A/cm^2)
$i_{(0(a))}$	Exchange current density at anode (A/cm^2)
$i_{(0(c))}$	Exchange current density at cathode (A/cm^2)
K_IC^i	Indentation fracture toughness (MPa·m$^{1/2}$)
M	Metal
min	Minutes
μm	Micrometre or micron
n	Number of electrons involved in the corrosion (electrochemical) reaction
nm	Nanometre (nm)
O^{2-}	Oxygen ions
Q_r	Reaction quotient
R	Universal gas constant (8.31 J/K-mol)
s	Seconds
T	Temperature (°C) or (in "K" unless stated otherwise)
T_a	Ageing temperature (°C) or (in "K" unless stated otherwise)
T_g	Glass transition temperature (°C) or (in "K" unless stated otherwise)
T_s	Transformation temperature (°C) or (in "K" unless stated otherwise)
TPa	Terra Pascal

Acronyms

A.C.	Alternating current (A)
AFM	Atomic force microscope
Al_2O_3	Aluminium oxide
AlO_4	Oxidoperoxy(oxo)alumane
ANN	Artificial neural network
AT	Alumina-titania
$BaSO_4$	Barium sulphate
BSCs	Bioinspired smart coatings
CA	Contact angle
Ca-P	Calcium phosphate
$CaSiO_3$	Calcium silicate
CCCs	Chromate conversion coatings
$[Ce(NO_3)]_3$	Cerium nitride
CP	Conductive polymer
CrO_3	Chromic acid
Cr_2O_3	Chromium oxide
CSSCs	Corrosion sensing smart coatings
CVD	Chemically vapour deposited
EB	Emeraldine base
e-coat	Electrodeposited coatings
Fe_2O_3	Ferrous oxide
Fe_2O_4	Ferrous dioxide
GNPs	Gold nanoparticles
H	Hydrogen
HA	Hydroxyapatite
HAP	Hazardous air pollutant
LB	Leucoemeraldine base
MEMS	Micro-electro-mechanical system/device
MgO	Magnesium oxide
MMC	Metal matrix composite
MS	Mild steel
NCs	Nanostructured coatings
NDS	Naphthalenedisulfonate
NEMS	Nano-electro-mechanical system/device
NNM	Nanoparticulate nanomaterials
OCP	Open circuit potential
PANI	Polyaniline
PB	Pernigraniline base
PMo_{12}	Molybdophosphate
PPy	Polypyrrole
PSSCs	Pressure sensing smart coatings

PTh	Polythiophene
SCs	Smart coating(s)
SEM	Scanning electron microscope
SHSCs	Self-healing smart coatings
SiO_2	Silicon dioxide
SiO_4	Silicate
SMA/Ps	Shape memory alloys/polymer
SME	Shape memory effect
SnO_2	Tin (IV) oxide (cassiterite)
SSCs	Smart switchable coatings
SMAT	Surface mechanical attrition treatment
TEM	Transmission electron microscope
TiO_2	Titanium dioxide
UV	Ultraviolet
UVCC	Ultraviolet curable coatings
VOC	Volatile organic compound
WSP	Water-soluble paints
YSZ	Yttria-stabilized zirconia
ZnO	Zinc oxide
$ZnSO_4$	Zinc sulphate

1 Smart Coatings
Science and Technology – From Fundamentals to Advances

1.1 INTRODUCTION

Metals and alloys, in commercial use, are prone to environmental degradation. The "coating" is a functional barrier, ranging from *nanometres* to *millimetres* in thickness, which protects the metallic structure from environmental degradation to some extent. It may be organic, inorganic, metallic, or polymeric but plays an important role in the proper functioning of the component or structure. In 1908, Henry Ford demonstrated the use of natural linseed resin oil, the world's first automobile protection coating, for his universal car Model-T. It was believed to be durable and economical, although the applying and drying time was inevitably longer. *DuPont*, USA used low-viscosity nitrocellulose-based compound to reduce drying time. A few years later, in 1924, General Motors adopted "Duco Finished" colour pigments and almost set a milestone for the automotive coating industry.

Besides nitrocellulose-based lacquer, which was highly productive until the 1940s, paint chemists wondered about the ultimate paint that gives a glossy and smooth surface. As a result, they invented some polymer-based binders for the paint, reflecting a comparatively glossier surface than the natural oil-based resin with increased productivity. The acrylic-based resin was used effectively as a topcoat till the 1970s. Meanwhile, automotive industries struggled to protect the major automotive parts from rusting. However, further advancements in coating industries resolved this issue using electrodeposited coatings (also known as "e-coat"), which protected the underlying metal by an anodic protection. With an ever-increasing demand for the improved efficiency, in the late 1990s, acrylic latex and polyurethane-dispersed water-borne coatings occupied almost 60% of the coating market. However, one of the downsides of water-borne coating is that it uses a volatile organic compound (VOC) such as petroleum-based solvents, which are hazardous from the environmental safety point of view. Conventionally, the prime focus of the industry was centred on the use of these coatings for protection and decoration purposes.

Over the last two decades, stringent environmental regulations have led the coating industries and paint chemists to look for alternatives. Significant progress in reduced VOC water-borne coatings, UV curable, and e-coat is noteworthy.

DOI: 10.1201/9781003200635-1

Further, voluminous efforts and studies have been devoted to understanding the scientific basis; controlling factors; improved functionality; and enhancement in mechanical, electrochemical, and wear properties for durability and longevity. Among the recent studies [1–10], the development of high-performance coatings with improved functionality with novel microstructures for drug delivery, transport, power, aerospace, and defence applications is notable. The technology at the driving seat of this 21st century is, of course, "nanotechnology" and has an obvious influence on the coating community. This further led to the development of thin nanocrystalline and nanostructured coatings for miniaturized devices such as micro-electro-mechanical systems (MEMS) and nano-electro-mechanical systems (NEMS). Of all these, piezo thin films, conducting polyaniline, (PANI) liquid crystals, and metal oxide thin films are quite popular and have been extensively used in MEMS and NEMS. For instance, the seminal work by Sazou and co-workers, Jadhav et al., Rohwerder et al., and Dhawan and co-workers [11–20] in the case of PANI- and polypyrrole (PPy)-based nanocomposite coatings have made significant contributions in controlling and protecting the corrosion of steel as well as in the MEMS applications.

In the current scenario, there is a growing need for commercial coating products with novel functions that can sense, interact, and respond to the environmental stimulus for protecting the underlying surface, often referred to as "Smart Coatings". The word "Smart Coatings" itself indicates that these coatings have to respond to stimuli whenever needed without compromising their passive characteristics. Therefore, the fundamental differences between conventional and smart coatings and some recent advances are explained below.

1.2 CONVENTIONAL COATINGS: SOLUTIONS AND ISSUES

As noted previously, the interaction between structural material and different environments plays a vital role, and the surface lies at the centre of the interaction. Therefore, strategies are adopted to protect the surface from the corrosive environments by the deposition of layer (relatively thinner than the dimensions of the bulk substrate). However, this so-called thin layer (of different materials), often known as "protective coatings", may be organic or inorganic, and ceramics should possess enough functional and structural properties to resist the environment. These typically contain high adhesion to the underlying substrate, thickness, strength, permeability, and importantly, coatings should act as a noble element during their service. Of all these, water-borne organic coatings have been preferred in many instances due to their inherent advantages, such as high adhesion, low toxicity (as it contains 80% water and small quantities of other solvents), and high surface coverage (compared to solvent-borne coatings).

Importantly, these coatings can act as an ideal primer with excellent heat and abrasion resistance. Conventional organic coatings protect the metal surface by the sacrificial anode (mostly zinc), barrier effect, and inhibition mechanisms. Mathiazhagan, Yuan, and Pan [21–23] have showcased the use of single-layer polymer–derived coatings for biosensor and corrosion protection applications.

These coatings are reported to exhibit excellent oxygen smearing-out and potential surface ennobling capabilities. Furthermore, Kouloumbi and Elsner [24,25] have produced multilayer thick epoxy resin-based organic coatings (thickness up to 400 μm) and demonstrated that the anticorrosive performance of the carbon steel could be enhanced (~5 times higher than that of the bare steel) without compromising structural properties and longevity. These studies also witnessed that the composition and orientation of primer and the size of polymeric particles significantly affect the performance of these coatings. It can be further construed that, to protect the metal surface, proper utilization of corrosion inhibitor and coating (or paint pigment) that can act as an impermeable barrier to corrosion species and moisture plays a vital role.

Besides this, conductive polymer (CP) coatings, including PANI, PPy, and polythiophene (PTh), are quite extensively used in the corrosion protection of metals. The CP can be synthesized *via* chemical and electrochemical means with electrical conductivity as an added advantage, making them suitable as an anticorrosive material [26–30]. The underlying corrosion protection mechanisms of CP can be stated as follows: (i) anodic protection mechanism, (ii) controlled inhibitor release mechanism, and (iii) doping of CP with some semiconductors. PANI coatings can be obtained from electronically conductive emeraldine salt (ES), widely accepted as an active corrosion control material. Further, leucoemeraldine base (LB), emeraldine base (EB), and pernigraniline base (PB) are some of the well-defined insulators of PANI. The PPy coatings have emerged as a new class of CP in the last two decades. Kowalski and co-workers [31] have produced single- and bi-layer PPy coatings on mild steel (MS) by an electrochemical deposition method. The inner and outer layers of PPy were doped with molybdophosphate (PMo_{12}) and phosphate (HPO_4^{2-}) ions and naphthalenedisulfonate (NDS) ions, respectively, resulting in thick coating (~5 μm). The inner layer doped with PMo_{12} and HPO_4^{2-} ions significantly stabilized the passive oxide film on MS substrate, while NDS in the outer layer prevented the decomposition of PMo_{12} and HPO_4^{2-} ions resulting in enhanced anticorrosive properties of MS. Several studies that have been made in the past also reported that the corrosion protection performance of PANI and PPy depends on the thickness of the layer, as thicker layers result in the improvement in anticorrosive performance [31–35]. Further, Kousik and Ocampo [36,37] have attempted to produce the multilayer PTh coatings on MS with the addition of acetonitrile and (0.2% w/w) of poly(3-decylthiophene-2,5-diyl). These coatings protected the MS surface by passivation due to the redox activity of PTh. The recent developments with nanostructure dopants such as nanowire, nanotube, nanorod, and nanofibres are noteworthy [38–50].

In the current scenario, a considerable number of efforts have been put to expand the design possibilities and alienate the intrinsic limits of the materials. For example, recent studies by Kathavate, Mirzakhanzadeh, and Li [51–58] have shown that zinc phosphate and zinc-aluminium coatings can effectively encounter corrosion issues in offshore and marine structures, thereby overcoming the inherent limits of the steel. The emanation of these studies revealed that nano TiO_2- and nano ZnO-incorporated phosphate coatings could significantly improve the corrosion resistance of steel (~8 times higher than that of the bare steel). Further, Al_2O_3-doped yttria-stabilized zirconia (YSZ) coatings deposited by plasma and thermal spray techniques can be used to protect the turbine blades made of nickel-based superalloys where the high-temperature oxidation often deteriorates the anticorrosive and mechanical properties of the turbine blades [59–61].

As mentioned earlier, one of the prima-facie objectives of coatings is to block the access of corrosive ions (electrolyte) to the underlying substrate. Although the conventional coatings have widened the inherent microstructural limits of metals, the practical applications of coatings demand reliability and longevity. There are several limitations with the direct applications of traditional coatings. For example, the operational temperature change may induce thermal degradation and chemical ageing, resulting in detachment and cracking in coatings [62–64]. Most of the polymer-based coatings lose their chemical stability and toughness due to cross-linking of polymer and change in chemical composition when deployed at high temperature and UV atmosphere, thereby leading to poor substrate adhesion [65,66]. Further, the physical polymer ageing may occur when polymer coatings are subjected to the temperature range below the glass transition temperature, T_g, thereby deteriorating the dielectric, mechanical, and thermal properties [67–70]. The use of such coatings below T_g establishes the non-equilibrium phase in polymer due to high enthalpy and volume incompatibilities (Figure 1.1), consequently reducing the glassy state. The physical ageing in polymer coatings triggers the coating contraction (Figure 1.2) though its effect at high-temperature equilibrium is insignificant. However, reheating of polymer above the T_g can reverse the physical ageing of polymer coatings.

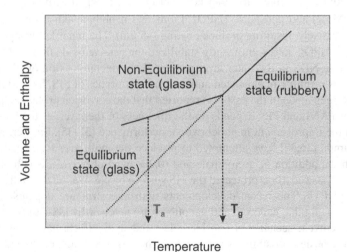

FIGURE 1.1 Variation in coating volume and enthalpy with respect to ageing temperature, T_a, and glass transition temperature, T_g, in polymer-based coatings. (Adopted and reproduced with permission from D.Y. Perera, *Progress in Organic Coatings*, 47 (2003) 61–66 [70].)

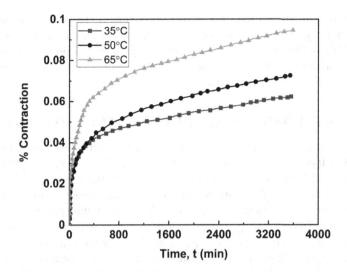

FIGURE 1.2   Variation in % coating contraction with respect to ageing time, *t* at different T_a. (Adopted and reproduced with permission from D.Y. Perera, *Progress in Organic Coatings*, 47 (2003) 61–66 [70].)

One interesting question to pose at this juncture is: How external electrolytes, ions, and gases penetrate into the coating? This question ultimately points towards the *intrinsic* and *extrinsic* defects associated with the coatings. The defects in the former category could arise due to polymer chemistry itself, while the latter is an outcome of external factors such as scratching and bubbling including pin-holes and thinned regions. Interestingly, the influence of extrinsic factors through intrinsic pathways deteriorates the structural integrity of the coatings. For example, the ununiform chemistry of the polymer resins may cause uneven (or unequal) cross-linking, which further creates localized pathways for the electrolyte ions to access the substrate. Due to CP's insoluble and non-fusible nature, their application on various metals is still a challenging task. Considerable efforts have been made to overcome these issues, which typically include chemical modifications and a well-designed functional group [10,71,72]. Surface functionalization *via* the addition of the sulfonate group as a polymer backbone in CP enhances the ion exchange properties, resulting in improved solubility [71,72]. Furthermore, blistering and bubbling in organic and inorganic coatings due to moisture absorption, moisture condensation, and unsuitable thinner may cause significant malfunctioning of the coating during service. Porosity and cathodic disbonding are the inherent defects associated with inorganic coatings such as nano ZnO/TiO_2-incorporated phosphate, Zn–Al polyphosphate, YSZ coatings, and Mg/Mn-phosphate. Besides this, several defects contribute to the structural disintegration of the coatings.

All the above observations highlight that these defects associated with conventional coatings may give rise to electrochemical corrosion on the metal surface, leading to the damage of the protective layer. This problem becomes even worse in biomedical implants, and drug delivery systems as the components/devices used in these systems are often subjected to the corrosive environment (i.e., body fluid or saliva). Therefore, there is an indeed requirement of a protective layer that can sense and give *stimuli* response to the change in the surrounding environment, often known as "Smart Coatings".

1.3 SMART COATINGS: A STATE-OF-THE-ART AND FUTURE

In today's "nano era", there is a continuous thurst to develop nanostructured coatings (NCs) that will exhibit the high strength–ductility synergy, superior wear resistance and lubricity, and corrosion resistance. At the same time, these NCs should act smarter and auto-responsive (or provide feedback) to the aggressive environment for enhanced longevity. Seen from this perspective, nano-architectured Smart Coatings (hereafter referred to as SCs) are the "intelligent coatings" that provide a stimuli response to the external (micro)environmental changes such as variation in pH, mechanical stresses, temperature (or heat induction), changes to the micro phobic nature of the surface, and most importantly, aggressive corrosion environment. Owing to their unique passive feature and ability to recover the functional performance when the need arises, sometimes they are also referred to as "environmental sensitive coatings" or "stimuli-responsive coatings". Therefore, benefitting from their pre-defined stimuli-responsive characteristics, SCs are designed to react spontaneously for diverse applications. These coatings can also extend the life of the structure and add additional functionalities to the existing coating systems. For instance, reversible thermochromic SCs can prevent the underlying surface of lightweight alloys (Mg and Al) and composites in a high-temperature environment and from aggressive corrosive ions [73]. Furthermore, self-healing SCs produced by incorporating urea-formaldehyde microcapsules into polymer reagent were turned out to be a major breakthrough among the coating community [74,75]. On this front, some popular existing SCs to date are defect sensing SCs, pressure sensing coatings, self-cleaning SCs, light or UV-sensing coatings, antifouling and antimicrobial SCs, electro-mechanical (or piezoelectric) SCs, etc. The profound review articles available on SCs to date revealed no single classification system for SCs [76,77]. Nevertheless, the classification of SCs may depend on the materials, inhibitors, response characteristics, and applications. One more important aspect for the classification of SCs is their functionality and end-users though none of the above coatings fit in an individual classification mentioned above.

As mentioned earlier, SCs can respond *via* triggering mechanisms from the external environment. However, this triggering response can be fully automatic or initiated by intelligent systems (such as sensors and actuators, often the SCs community calls them piezoelectric coatings). The SCs community would prefer automatic responses in actual applications as it eliminates the additional need for external triggers, thereby reducing production/manufacturing costs. The automatic triggers are based on the intrinsic or inherent functional features of the materials, while external trigger mediated response generally depends on the working nature, type, and intervention of external agents. There are focused studies in the literature indicating the automatic response of SCs. However, the commensurate studies which use external triggering mechanisms in SCs are less in number (or such types of coatings are not fully developed). The schematic in Figure 1.3 represents the automatic stimuli response of SCs. This illustrates that the degradation *via* UV rays is responsible for the delamination and swelling in the topcoat and primer on the substrate. However, at the same time, UV-degradation also promotes the release of healing agents (mostly polymer-based and green inhibitors) and prevents damage to the coatings. These common but important mechanisms of stimuli response in SCs are explored below.

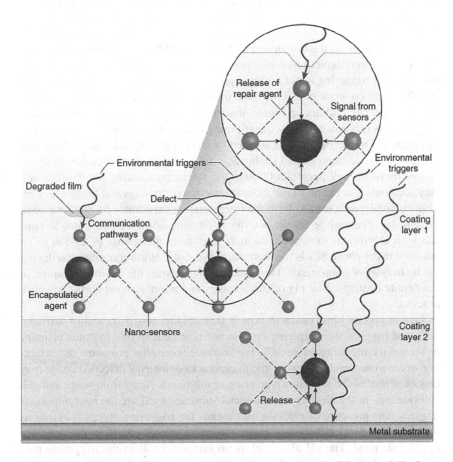

FIGURE 1.3 Schematic illustration of triggering mechanisms in SCs (especially from environmental triggers). (Adopted and reprinted with permission from A.S.H. Makhlouf, Woodhead Publishing Series in Materials Science and Engineering, Elsevier (2014) [76].)

1.4 MECHANISMS OF STIMULI RESPONSE

In this section, different mechanisms of stimuli response (also called triggering mechanisms) are outlined. As explained in the above section, SCs in their applications may respond by automatic or external intelligent agents; therefore, various triggering mechanisms can be proposed for the above stimuli response. This eventually includes chemical, electrochemical, physical, and mechanical triggering. These triggering mechanisms could result from macroscopic and microscopic changes (damages) in the coatings.

1.4.1 CHEMICAL STIMULI RESPONSE

Chemical triggering is perhaps the most basic and commonly used technique in SCs operated *via* water-soluble inhibitors. However, in most of the applications, there is less control over the water level in SCs. The observations of Scholes et al. [78] and Prosek et al. [79] have indicated that the release of chromate-inhibited primers continues till the immersion of the material/SCs. Their observations also revealed that upon depletion of inhibitor due to complete wetness of the coatings, the chemical triggering is not an appropriate mechanism. Furthermore, Galvele and Dias and co-workers [80,81] have introduced a Ce-impregnated Zeolite-X SCs with low pH inhibitor and found that anodic attack is more predominant due to hydrolysis reaction and subsequent pitting occurs due to acidification. During this chemical triggering process adopted by Dias et al. [81] (schematically shown in Figure 1.4), Ce-impregnated zeolite SCs are produced on an aluminium sheet.

The moisture will penetrate through the microcrack during the damage and may accumulate in the SCs both from the surface of coatings and the crack edge. This eventually will lead to form the intermetallics favourable for the formation of cathodic areas, while anodic activity is obvious as the defects come in contact with aluminium substrate. The main reason for the release of Ce-ion as an inhibitor in the above SCs is the increase in hydrogen, H ion concentration during the hydrolysis of aluminium. This eventually promotes the Ce-ion exchange in Ce-Zeolite coatings. The pH of the coatings and water also supports the entire process.

One important observation to note is that despite environmental or artificial mediated triggers, the triggering process needs to be activated by some primary or secondary mechanisms in SCs. For instance, corrosion promotes the activation of environmental triggers, while microcracks or internal defects/damage may endorse the artificial mediated triggering of inhibitors. Apart from water and pH, chloride ions in the electrolyte and metal/substrate itself are the next important triggers. The presence of chloride ions makes the triggering process obvious as the corrosion process becomes easier (or rather initiates itself in the presence of chloride ions). The pH always plays an important role in the triggering process (as a whole, an individual or a mediator) and is therefore often recognized as a pseudo-metal ion trigger [76]. There is a reason to believe this argument because an increase in the concentration of H ions during the hydrolysis of metal causes local acidification, and pH may help additionally to trigger the inhibitors.

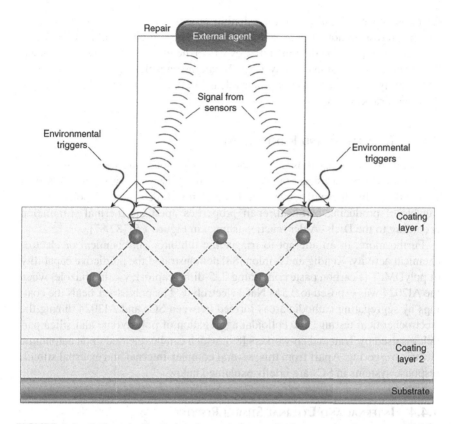

FIGURE 1.4   Schematic displaying the environmental trigger assisted response of SCs (methodology proposed by Dias et al. [81]). (Adopted and reprinted with permission from A.S.H. Makhlouf, Woodhead Publishing Series in Materials Science and Engineering, Elsevier (2014) [76].)

Although the chemical triggers are of great importance to the SCs community, the damage due to mechanical and related impacts cannot be healed by environmentally assisted triggers. Therefore, in the next section, the emphasis is given to mechanical stimuli response mechanisms of SCs.

1.4.2   MECHANICAL STIMULI RESPONSE

Mechanical triggering is also one of the important stimuli response mechanisms in SCs and can be used for triggering the inhibitors in polymers and composites. The mechanical triggering has been extensively studied by Fischer, Dry, and White and co-workers [82–84]. They demonstrated that reinforced polymer matrix composites could form passive and smart-self responsive/repair layers triggered by dynamic mechanical loading. These composites consist of (i) an agent/stimulus which releases the chemicals on external mechanical damage, (ii) a chemical monomer inside a fibre that is capable of self-repairing the damaged part, and (iii) the hardening process *via* cross-linking polymer. On an interesting note, the mixed-mode application of chemical and mechanical stimuli response is demonstrated by Miccichè et al. [85]. They have shown that mechanical loading can physically break the capsules of clay particles, and additionally, water can promote their release in the coatings to heal/self-repair the microcrack or defects.

1.4.3 TEMPERATURE AND REDOX ACTIVITY

Temperature can be used as one of the stimuli response mechanisms in SCs in the case of alloys with shape memory effects and polymers (with typically mixed T_g) [82,86]. The thermally reversible bonds (cross-links) in some polymers are capable of producing self-heal/repair properties upon the thermal stimulation according to the Diels–Alder reaction (shown in Figure 1.5) [82,87].

Furthermore, in an attempt to trigger the inhibitor *via* chemical or electrochemical activity, Kendig and Kinlon [88] demonstrated the predictive capability of polyDMcT (a carbon paste containing 2, 5-dimercapto-1,3,4-thiadiazole) when the Al2024 was exposed to 0.5 M NaCl electrolyte. The polyDMcT heals the coatings by segregating cathodic areas formed between SCs and Al2024 during the electrochemical testing. The colloidal aggregation of polystyrene and silica particles is used by Trau and co-workers [89], which can be released upon autonomic electric triggering. Apart from this, several complex internal and external stimuli response systems in SCs are briefly explained below.

1.4.4 INTERNAL AND EXTERNAL STIMULI RESPONSE

These types of triggering mechanisms are gaining a wide response from space, defence, and security organizations and are being extensively developed nowadays. Most of these triggering mechanisms contain the materials like optoelectronics and piezoelectrics, which can exhibit stimuli response to external changes. These materials can sense the changes in SCs and metal-coating interfaces and then trigger the inhibitors, followed by response initiation. The typical applications of these stimuli responses are obvious in transport and power industries. The efforts are being made to develop such triggering inhibitors that can sense the environmental changes occurring during power transportation and avoid the surface icing by

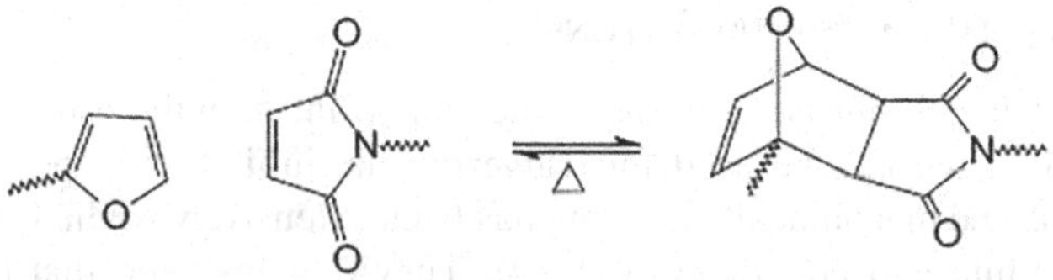

FIGURE 1.5 Schematics representing Diels–Alder reaction in cross-linked polymers upon thermal stimulation. (Adopted and reprinted with permission from H.R. Fischer, *Natural Science*, 2 (2010) 873–901 [82].)

releasing anti-freeze liquid. The two terminologies, "hydrophilic and hydrophobic" surfaces, may play a vital role in designing such triggering mechanisms.

To sum up, a series of stimuli-responsive materials/inhibitors are being developed and are under evaluation. The efforts are also being made to produce such inhibitors in a less-expensive manner with real-time monitoring/assessment of data (i.e., using sensors) rather than predicting the time-consuming life cycle analysis based on historical data.

1.5 ADVANCEMENT IN SCs AND CONCLUSIONS

One important question to pose at the end of this chapter is: "Is it possible to produce toxin-free inhibitor SCs on the bulk scale (for mesoscale applications)?" The future trends in SCs are hidden in the answers to this question. For instance, removing toxic chromate and fluoride from SCs is an important open problem for the SCs community. To date, the coating industries are continuously starving to find the alternative for the above inhibitors but far from complete (since chromate possesses exceptional characteristics). In one such attempt, efforts are directed to produce the chromate-free inhibitor with a synergistic combination of chemical and electrochemical stimuli response using a series of chemicals. At present, coatings industries are also looking forward to develop the SCs with multiple functionalities (which are capable of energy generation and odour absorption and can act as an insulator). Furthermore, the profound review of the prior literature also indicates the possible scope for the development of fixed and mobile coatings in a cost-effective manner in future for transport, space, and defence applications.

In summary, the present chapter provides an overview of different conventional coatings and SCs with special emphasis on their applying techniques and functional properties. A comprehensive review on the development and advancement of SCs is presented with the aid of commercial applications, which further reveals the various stimuli response and healing/repair mechanisms. Some interesting observations from the previous literature may be helpful for the SCs community to develop the coatings which can sense the data/changes and automatically trigger the inhibitor. Such mechanisms would be beneficial in designing the SCs for transport and power industries. The challenge of creating SCs for high-temperature applications in aerospace and defence instruments is an important open problem and needs a comprehensive approach in future.

REFERENCES

1. M.A. Quraishi, A. Singh, V.K. Singh, D.K. Yadav, A.K. Singh, Green approach to corrosion inhibition of mild steel in hydrochloric acid and sulphuric acid solutions by the extract of Murraya koenigii leaves, *Materials Chemistry and Physics*, 122 (2010) 114–122.
2. E. Sharmin, O. Rahman, F. Zafar, D. Akram, M. Alam, S. Ahmad, Linseed oil polyol/ZnO bionanocomposite towards mechanically robust, thermally stable, hydrophobic coatings: a novel synergistic approach utilizing a sustainable resource, *RSC Advances*, 5 (2015) 47928–47944.

3. J. Buchweishaija, Phytochemicals as green corrosion inhibitors in various corrosive media: a review, *Tanzania Journal of Science*, 35 (2009) 77–92.

4. K.P. Vinod Kumar, M. Sankara Narayanan Pillai, G. Rexin Thusnavis, Green corrosion inhibitor from seed extract of Areca catechu for mild steel in hydrochloric acid medium, *Journal of Materials Science*, 46 (2011) 5208–5215.

5. M. Izadi, T. Shahrabi, B. Ramezanzadeh, Active corrosion protection performance of an epoxy coating applied on the mild steel modified with an eco-friendly sol-gel film impregnated with green corrosion inhibitor loaded nanocontainers, *Applied Surface Science*, 440 (2018) 491–505.

6. R. Otero, A.L. Vázquez de Parga, J.M. Gallego, Electronic, structural and chemical effects of charge-transfer at organic/inorganic interfaces, *Surface Science Reports*, 72 (2017) 105–145.

7. K.E.A. AbouAitah, A.A. Farghali, Mesoporous silica materials in drug delivery system: pH/glutathione-responsive release of poorly water-soluble pro-drug querecetin from two and three-dimensional pore-structure nanoparticles, *Journal of Nanomedical and Nanotechnology*, 7 (2016) 1000360.

8. T.H. Tran, A. Vimalanandan, G. Genchev, J. Fickert, K. Landfester, D. Crespy, M. Rohwerder, Regenerative nano-hybrid coating tailored for autonomous corrosion protection, *Advanced Materials*, 27 (2015) 3825–3828.

9. S.K. Natarajan, S. Selvaraj, Mesoporous silica nanoparticles: importance of surface modifications and its role in drug delivery, *RSC Advances*, 4 (2014) 14328.

10. X. Ma, L. Xu, W. Wang, Z. Lin, X. Li, Synthesis and characterization of composite nanoparticles of mesoporous silica loaded with inhibitor for corrosion protection of Cu-Zn alloy, *Corrosion Science*, 120 (2017) 139–147.

11. D. Sazou, C. Georgolios, Formation of conducting polyaniline coatings on iron surfaces by electropolymerization of aniline in aqueous solutions, *Journal of Electroanalytical Chemistry*, 429(1–2) (1997) 81–93.

12. N. Jadhav, C.A. Vetter, V.J. Gelling, The effect of polymer morphology on the performance of a corrosion inhibiting polypyrrole/aluminum flake composite pigment, *Electrochimica Acta*, 102 (2013) 28–43.

13. H. Bhandari, S.A. Kumar, S.K. Dhawan, Conducting polymer nanocomposites for anticorrosive and antistatic applications, In: Ebrahimi, F (ed.) *Nanocomposites: New Trends and Developments*, pp. 73–96. InTech Open, Rijeka (2012).

14. D. Sazou, Electrodeposition of ring-substituted polyanilines on Fe surfaces from aqueous oxalic acid solutions and corrosion protection of Fe, *Synthetic Materials*, 118(1–3) (2001) 133–147.

15. D. Sazou, M. Kourouzidou, E. Pavlidou, Potentiodynamic and potentiostatic deposition of polyaniline on stainless steel: electrochemical and structural studies for a potential application to corrosion control, *Electrochimica Acta*, 52(13) (2007) 4385–4397.

16. D. Sazou, M. Kourouzidou, Electrochemical synthesis and anticorrosive properties of nafion®-poly(aniline-coo-aminophenol) coatings on stainless steel, *Electrochimica Acta*, 54(9) (2009) 2425–2433.

17. D. Kosseoglou, R. Kokkinofta, D. Sazou, FTIR spectroscopic characterization of nafion®-polyaniline composite films employed for the corrosion control of stainless steel, *Journal Solid State Electrochemical*, 15(11–12) (2011) 2619–2631.

18. D. Sazou, D. Kosseoglou, Corrosion inhibition by nafion®-polyaniline composite films deposited on stainless steel in a two-step process, *Electrochimica Acta*, 51(12) (2006) 2503–2511.

19. M. Rohwerder, A. Michalik, Conducting polymers for corrosion protection: what makes the difference between failure and success? *Electrochimica Acta*, 53(3) (2007) 1300–1313.

20. H. Bhandari, V. Choudhary, S.K. Dhawan, Influence of self-doped poly(aniline-co-4-amino-3-hydroxy-naphthalene-1-sulfonic acid) on corrosion inhibition behaviour of iron in acidic medium, *Synthetic Materials*, 161(9–10) (2011) 753–762.
21. A. Mathiazhagan, R. Joseph, Nanotechnology – a new prospective in organic coating, *International Journal of Chemical Engineering and Applications*, 2(4) (2011) 225–237.
22. Y.C. Yuan, T. Yin, M.Z. Rong, M.Q. Zhang, Self-healing in polymers and polymer composites. Concepts, realization and outlook: a review, *Polymer Letters*, 2(4) (2008) 238–250.
23. T. Pan, Intrinsically conducting polymer-based heavy duty and environmentally friendly coating system for corrosion protection of structural steels, *Spectroscopy Letters*, 46 (2013) 268–276.
24. N. Kouloumbi, P. Moundoulas, Anticorrosive performance of organic coatings on steel surfaces exposed to deionized water, *Pigment and Resin Technology*, 31(4) (2002) 206–215.
25. C.I. Elsner, E. Cavalcanti, O. Ferraz, A.R. Di Sarli, Evaluation of the surface treatment effect on the anticorrosive performance of paint systems on steel, *Progress in Organic Coatings*, 48 (2003) 50–62.
26. D. Tallman, G. Spinks, A. Dominis, G. Wallace, Electroactive conducting polymers for corrosion control, *Journal of Solid State Electrochemistry*, 6(2) (2002) 73–84.
27. G. Spinks, A. Dominis, G. Wallace, D. Tallman, Electroactive conducting polymers for corrosion control, *Journal of Solid State Electrochemistry*, 6 (2) (2002) 85–100.
28. P. Zarras, J.D. Stenger-Smith, Y. Wei, *Electroactive Polymers for Corrosion Control*, American Chemical Society, Washington, DC, (2003).
29. M. Rohwerder, Conducting polymers for corrosion protection: a review, *International Journal of Materials Research*, 100(10) (2009) 1331–1342.
30. P.P. Deshpande, N.G. Jadhav, V.J. Gelling, D. Sazou, Conducting polymers for corrosion protection, *Journal of Coatings Technology and Research*, 11(4) (2014) 473–494.
31. D. Kowalski, M. Ueda, T. Ohtsuka, Corrosion protection of steel by bi-layered polypyrrole doped with molybdophosphate and naphthalenedisulfonate anions, *Corrosion Science*, 49(3) (2007) 1635–1644.
32. G. Bereket, E. Hür, The corrosion protection of mild steel by single layered polypyrrole and multilayered polypyrrole/poly(5-amino-1-naphthol) coatings, *Progress in Organic Coatings*, 65(1) (2009) 116–124.
33. W. Su, J.O. Iroh, Electrodeposition mechanism, adhesion and corrosion performance of polypyrrole and poly(nmethylpyrrole) coatings on steel substrates, *Synthetic Metals*, 114(3) (2000) 225–234.
34. P. Herrasti, AI del Rio, J. Recio, Electrodeposition of homogeneous and adherent polypyrrole on copper for corrosion protection, *Electrochimica Acta*, 52(23) (2007) 6496–6501.
35. Y.F. Jiang, X.W. Guo, Y.H. Wei, C.Q. Zhai, W.J. Ding, Corrosion protection of polypyrrole electrodeposited on AZ91 magnesium alloys in alkaline solutions, *Synthetic Metals*, 139(2) (2003) 335–339.
36. G. Kousik, S. Pitchumani, N.G. Renganathan, Electrochemical characterization of polythiophene-coated steel, *Progress in Organic Coatings*, 43(4) (2001) 286–291.
37. C. Ocampo, E. Armelin, F. Liesa, C. Alemán, X. Ramis, J.I. Iribarren, Application of a polythiophene derivative as anticorrosive additive for paints, *Progress in Organic Coatings*, 53(3) (2005) 217–224.
38. Y.Z. Long, M.M. Li, C. Gu, M. Wan, J.L. Duvail, Z. Liu, Z. Fan, Recent advances in synthesis, physical properties and applications of conducting polymer nanotubes and nanofibers, *Progress in Polymer Science*, 36(10) (2011) 1415–1442.

39. Y. Guo, Y. Zhou, Polyaniline nanofibers fabricated by electrochemical polymerization: a mechanistic study, *European Polymer Journal*, 43(6) (2007) 2292–2297.

40. J. Huang, R.B. Kaner, Nanofiber formation in the chemical polymerization of aniline: a mechanistic study, *Angewandte Chemie International Edition*, 43(43) (2004) 5817–5821.

41. J. Stejskal, I. Sapurina, M. Trchová, Polyaniline nanostructures and the role of aniline oligomers in their formation, *Progress in Polymer Science*, 35(12) (2010) 1420–1481.

42. M. Woodson, J. Liu, Guided growth of nanoscale conducting polymer structures on surface-functionalized nanopatterns, *Journal of the American Chemical Society*, 128(11) (2006) 3760–3763.

43. D. Li, J. Huang, R.B. Kaner, Polyaniline nanofibers: a unique polymer nanostructure for versatile applications, *Accounts of Chemical Research*, 42(1) (2008) 135–145.

44. C. Ge, X. Yang, B. Hou, Synthesis of polyaniline nanofiber and anticorrosion property of polyaniline epoxy composite coating for Q235 steel, *Journal of Coatings Technology and Research*, 9(1) (2012) 59–69.

45. N. Elhalawany, M.A. Mossad, M.K. Zahran, Novel water based coatings containing some conducting polymers nanoparticles (CPNs) as corrosion inhibitors, *Progress in Organic Coatings*, 77(3) (2014) 725–732.

46. P. Deshpande, S. Vathare, S. Vagge, E. Tomšík, E.J. Stejskal, Conducting polyaniline/multi-wall carbon nanotubes composite paints on low carbon steel for corrosion protection: electrochemical investigations, *Chemical Papers*, 67(8) (2013) 1072–1078.

47. B. Cho, H. Lim, H-N. Lee, Y.M. Park, H. Kim, H-J. Kim, High-capacity and cycling-stable polypyrrole-coated MWCNT@polyimide core-shell nanowire anode for aqueous rechargeable sodium-ion battery, *Surface and Coatings Technology*, 407 (2021) 126797.

48. C. Wang, X. Li, C. Tong, A. Cai, H. Guo, H. Yin, Preparation, invitro bioactivity and osteoblast cell response of $Ca-Ta_2O_5$ nanorods on tantalum, *Surface and Coatings Technology*, 391 (2020) 195701.

49. A. Claypole, J. Claypole, T. Claypole, D. Gethin, L. Kilduff, The effect of plasma functionalization on the print performance and time stability of graphite nanoplatelet electrically conducting inks, *Journal of Coatings Technology and Research* (2020). https://doi.org/10.1007/s11998-020-00414-4.

50. C. Garcia-Cabezon, C. Salvo-Comino, C. Garcia-Hernandez, M.L. Rodriguez-Mendez, F. Martin-Pedrosa, Nanocomposites of conductive polymers and nanoparticles deposited on porous material as a strategy to improve its corrosion resistance, *Surface and Coatings Technology*, 403 (2020) 126395.

51. N.S. Bagal, V.S. Kathavate, P.P. Deshpande, Nano TiO_2 phosphate conversant coatings – a chemical approach, *Electrochemical Energy Technology*, De Gruyter, 4 (2018) 47–54.

52. V.S. Kathavate, N.S. Bagal, P.P. Deshpande, Corrosion protection performance of nano TiO_2 incorporated phosphate coatings obtained by anodic electrochemical treatment, *Corrosion Reviews*, 37(6) (2019) 565–578.

53. V.S. Kathavate, D.N. Pawar, N.S. Bagal, P.P. Deshpande, Role of nano ZnO particles in the electrodeposition and growth mechanism of phosphate coatings for enhancing the anti-corrosive performance of low carbon steel in 3.5% NaCl aqueous solution, *Journal of Alloys and Compounds*, 823 (2020) 153812.

54. V.S. Kathavate, P.P. Deshpande, Role of nano TiO_2 and nano ZnO particles on enhancing the electrochemical and mechanical properties of electrochemically deposited phosphate coatings, *Surface and Coatings Technology*, 394 (2020) 125902.

55. J.O. Berghaus, M. Boulos, J. Brogan, A.C. Bourtsalas, A. Dolatabadi, M. Dorfman, et al., The 2016 thermal spray roadmap, *Journal of Thermal Spray Technology*, 25(8) (2016) 1376–1440.

56. N. LeBozec, D. Thierry, D. Persson, J. Stoulil, Atmospheric corrosion of zinc-aluminum alloyed coated steel in depleted carbon dioxide environments, *Journal of the Electrochemical Society*, 165(7) (2018) C343–C353.

57. Z. Mirzakhanzadeh, A. Kosari, M.H. Moayed, R. Naderi, P. Taheri, J.M.C. Mol, Enhanced corrosion protection of mild steel by the synergetic effect of zinc aluminum polyphosphate and 2-mercaptobenzimidazole inhibitors incorporated in epoxy-polyamide coatings, *Corrosion Science*, 138 (2018) 372–379.

58. W. Li, L. Shi, J. Zhang, J. Cheng, X. Wang, Double-layered surface decoration of flaky aluminum pigments with zinc aluminum phosphate and phytic acid–aluminum complexes for high-performance waterborne coatings, *Powder Technology*, 362 (2020) 462–473.

59. L. Gao, H. Guoa, L. Wei, C. Li, S. Gong, H. Xu, Microstructure and mechanical properties of yttria stabilized zirconia coatings prepared by plasma spray physical vapor deposition, *Ceramics International*, 41 (2015) 8305–8311.

60. X. Song, Z. Liu, M. Kong, C. Lin, L. Huang, X. Zheng, Y. Zeng, Thermal stability of yttria-stabilized zirconia (YSZ) and YSZ-Al$_2$O$_3$ coatings, *Ceramics International*, 43(2017) 14321–14325.

61. X. Luo, Z. Ning, L. Zhang, R. Lin, H. He, J. Yanga, Y. Yang, J. Liao, N. Liu, Influence of Al$_2$O$_3$ overlay on corrosion resistance of plasma sprayed yttria stabilized zirconia coating in NaCl-KCl molten salt, *Surface and Coatings Technology*, 361 (2019) 432–437.

62. D.R. Bauer, D.F. Mielewski, J.L. Gerlock, Photooxidation kinetics in crosslinked polymer coatings, *Polymer Degradation and Stability*, 38 (1992) 57–67.

63. D.F. Mielewski, D.R. Bauer, J.L. Gerlock, Determination of hydroperoxide concentrations in crosslinked polymer coatings containing hindered amine light stabilizers, *Polymer Degradation and Stability*, 41 (1993) 323–331.

64. D.R. Bauer, Application of failure models for predicting weatherability in automotive coatings, *ACS Symposium Series*, 722 (1999) 378–395.

65. D.Y. Perera, M. Oosterbroek, Hygrothermal stress evolution during weathering in organic coatings, *Journal of Coatings Technology*, 66 (1994) 83–88.

66. D.Y. Perera, P. Schutyser, C. De Lame, E.D. Vanden, On film formation and physical aging in organic coatings, *Polymer Materials Science & Engineering*, 73 (1995) 187–188.

67. L.C.E. Struik, *Physical Aging in Amorphous Polymers and Other Materials*, Elsevier, Amsterdam, 1978.

68. G.B. McKenna, Glass formation and glassy behavior, In: Booth, C., Price, C. (eds.) *Polymer Properties – Comprehensive Polymer Science*, vol. II, pp. 311–322. Pergamon Press, Oxford (1989).

69. R.C. Warren, The effect of aging and annealing on the physical properties of nitrocellulose plasticized with nitroglycerine, *Polymer*, 31 (1990) 861–868.

70. D.Y. Perera, Physical ageing of organic coatings, *Progress in Organic Coatings*, 47 (2003) 61–76.

71. J.R. Santos Jr, L.H.C. Mattoso, A.J. Motheo, Investigation of corrosion protection of steel by polyaniline films, *Electrochimica Acta*, 43(3–4) (1998) 309–313.

72. Y. Şahin, K. Pekmez, A. Yıldız, Electrochemical preparation of soluble sulfonated polymers and aniline copolymers of aniline sulfonic acids in dimethylsulfoxide, *Journal of Applied Polymer Science*, 90(8) (2003) 2163–2169.

73. S.B. Ulaeto, J.K. Pancrecious, T.P.D. Rajan, B.C. Pai, Smart coatings, In: *Noble Metal-Metal Oxide Hybrid Nanoparticles*, pp. 341–372. Elsevier (2019). https://doi.org/10.1016/B978-0-12-814134-2.00017-6.

74. S.R. White, N.R. Sottos, P.H. Geubelle, J.S. Moore, M.S. Kessler, S.R. Sriram, E.N. Brown, S. Viswanathan, Autonomic healing of polymer composites, *Nature*, 409 (2001) 494–497.

75. E.N. Brown, M.S. Kessler, N.R. Sottos, S.R. White, In situ poly(urea-formaldehyde) microencapsulation of dicyclopentadiene, *Journal of Microencapsulation*, 20 (2003) 719–730.

76. A.S.H. Makhlouf, *Handbook of Smart Coatings for Materials Protection*, Woodhead Publishing Series in Metals and Surface Engineering, Elsevier, Cambridge, 2014.

77. J. Baghdachi, *Smart Coatings*, ACS Symposium Series, American Chemical Society, Washington, DC, 2009.

78. F.H. Scholes, S.A. Furman, A.E. Hughes, T. Nikpour, N. Wright, P.R. Curtis, C.M. Macrae, S. Intem, A.J. Hill, Chromate leaching from inhibited primers – Part I characterization of leaching, *Progress in Organic Coatings*, 56(1) (2006) 23–32.

79. T. Prosek, D. Thierry, A model for the release of chromate from organic coatings, *Progress in Organic Coatings*, 49(3) (2004) 209–217.

80. J.R. Galvele, Transport processes and mechanism of pitting of metals, *Journal of the Electrochemical Society*, 123(4) (1976) 464–474.

81. S.A.S. Dias, S.V. Lamaka, C.A. Nogueira, T.C. Diamantino, M.G.S. Ferreira, Sol–gel coatings modified with zeolite fillers for active corrosion protection of AA2024, *Corrosion Science*, 62 (2012) 153–162.

82. H.R. Fischer, Self repairing systems – a dream or reality, *Natural Science*, 2 (2010) 873–901.

83. C. Dry, Procedures developed for self-repair of polymer matrix composite materials, *Composite Structures*, 35(3) (1996) 263–269.

84. S.R. White, N.R. Sottos, P.H. Geubelle, J.S. Moore, M.R. Kessler, S.R. Sriram, E.N. Brown, S. Viswanathan, Autonomic healing of polymer composites, *Nature*, 409 (2001) 794–797.

85. F. Miccichè, H.R. Fischer, R. Varley, S. van der Zwaag, Moisture induced crack filling in barrier coatings containing montmorillonite as an expandable phase, *Surface and Coatings Technology*, 202 (2008) 3346–3353.

86. Y. Gonzalez-Garcia, J.M.C. Mol, T. Muselle, I. De Graeve, G. Van Assche, G. Scheltjens, B. Van Mele, H. Terryn, A combined mechanical, microscopic and local electrochemical evaluation of self-healing properties of shape-memory polyurethane coatings, *Electrochimica Acta*, 56(26) (2011) 9619–9626.

87. O. Diels, K. Alder, Synthesen in der hydroa-romatischen Reihe, *Liebigs Annalen der Chemie*, 460(1) (1928) 98–122.

88. M. Kendig, P. Kinlen, Demonstration of galvanically stimulated release of a corrosion inhibitor – basis for 'smart' corrosion inhibiting materials, *Journal of the Electrochemical Society*, 154(4) (2007) C195–C201.

89. M. Trau, D.A. Saville, I.A. Aksay, Filed-induced layering of colloidal crystals, *Science*, 272 (1996) 706–709.

INTERNET SOURCE

1. D.M. Lamb, The birth of automotive coatings, *Coatings Technology*, 14(02) (2017).

2 Smart Coatings for Corrosion Protection

2.1 INTRODUCTION

"What is the supreme gift to the human society?" Of course, metals and their alloys. Owing to their outstanding functional and engineering properties, various metals and their alloys have served the emerging needs of human life that changes from time to time for better sustainability. The immediate question to pose at this juncture from the application point of view of metals is: "Which properties do any material scientist need to consider?" The answer to this question may reflect in mechanical, physical, and chemical senses. The first two properties are generally expressed in terms of constants, while chemical properties are explicitly dependent on the precise environmental (actual working) conditions during service. In line with this, one of the downsides of metals (or their alloys) is that they are thermodynamically unstable in their refined form (highest free energy and where chemical processing/properties are involved) and tend to corrode in their environment. Therefore, in general, corrosion can be defined as a chemical or electrochemical reaction between metal or an alloy and its environment [1]. The immediate consequences of the corrosion process in any material are the deterioration in engineering properties such as strength, ductility, hardness, and fracture toughness. From an electrochemical society point of view, corrosion involves the transfer of charge (electrons) between the metal electrode and chemical species. The process of corrosion in any material results in its decrease in free energy, ΔG, and can be related to the potential of corroding material according to

$$\Delta G = -nE_mF \qquad (2.1)$$

Here, n and E_m represent the number of electrons involved in the reaction and potential of corroding material, respectively, while F specifies Faraday's constant (96,500 C/mol). The typical causes of corrosion on the metal surface are mostly heterogeneity, lower oxygen concentration, stress concentration, hydrogen embrittlement, and contact between two dissimilar surfaces [2]. Over the years, numerous corrosion prevention strategies have been developed to prevent the deterioration of the metal surface. On this front, surface modification by alloying, surface treatments such as shot peening and surface mechanical attrition treatment, applying primer or coatings, and modification of the metal environment by inhibitors have been extensively used. Since one of the objectives of the present chapter is to get insights into the applications of smart coatings (SCs) in corrosion and antifouling prevention, therefore understanding the fundamentals of

DOI: 10.1201/9781003200635-2

corrosion and its kinetics is of paramount importance from the development of coatings (and eventually the surface protection) viewpoint. In view of this, we focus on some common but important electrochemical aspects of corrosion in the below section. The tedious mathematical formulation part in corrosion kinetics is not mentioned, thereby presenting the information vividly and concisely.

2.2 ELECTROCHEMICAL ASPECTS OF CORROSION: FUNDAMENTAL VIEWPOINT

As stated earlier, the corrosion process has two viewpoints: (i) chemical and (ii) electrochemical. Understanding the corrosion process and its kinetics in an electrochemical sense require the basics of the electrode (generally metal or alloys), electrolyte (solvated solution), charge transfer, electrode potential, and polarization. The fundamental concept of metal lattice states that positively charged ions sit at the fixed position, while negatively charged electrons move freely inside the metal lattice. On the other hand, aqueous electrolyte consists of various species, i.e., cations, anions, water dipoles, and trace impurities, that are randomly oriented (freely moving). During the immersion of metal electrode into an aqueous electrolyte, transfer of ions takes place *via* a simultaneous transport of ions through an electrolyte, accumulation of electrons within the metal electrode, and transfer of electrons at the metal electrode/electrolyte interface constituting an electrochemical reaction (cell) in the following manner: (i) positively charged ions from metal electrode have a tendency to move in aqueous electrolyte leaving behind the negatively charged electrons and (ii) the dissolved metal ions have a marked tendency to get deposited on the electrode surface. The immediate consequence of the former one is the oxidation of the metal to its own ions according to oxidation reaction (equation 2.2), while the latter one indicates the reduction reaction (equation 2.3) as a result of dissolution and deposition of metal ions on the electrode surface. Furthermore, the accumulation of negative charge on the electrode results in an increase in the potential difference between electrode and electrolyte:

$$M \rightarrow M^{z+} + ze^- \qquad (2.2)$$

$$M^{z+} + ze^- \rightarrow M \qquad (2.3)$$

Equation 2.2 represents the anodic reaction that results in oxidation of the metal electrode, while the reaction in equation 2.3 specifies the cathodic portion of the electrochemical cell reflecting the reduction of metal ions in an electrolyte and their subsequent deposition at the metal surface. The potential at which the rate of cathodic and anodic reactions is same (or equal) is known as equilibrium potential and can be evaluated by the Nernst equation [3] presented below:

$$E = E_0 \pm \frac{RT}{nF} \ln Q_r \qquad (2.4)$$

where E_0 is the standard electrode potential, R represents the universal gas constant (8.31 J/K-mol), T specifies the temperature in Kelvin (K), n is the number of electrons transferred, and F is Faraday's constant (96,500 C/mol). The term Q_r in the above equation is indicative of reaction quotient, a ratio of chemical activities of reactants and products from oxidation and reduction reactions. The positive sign in equation 2.4 represents the reduction potential, while the negative sign indicates the oxidation potential. It can be inferred from the Nernst equation that the equilibrium potential has a marked dependence on standard electrode potential, temperature, and ionic activity.

The tendency of metal dissolution (or forming metal ions) in electrolyte depends on E_0. Therefore, metal will corrode at a faster rate if the value of E_0 is negative. Note that E_0 is referred to as standard reduction potential in many instances [4]. The values of E_0 for different metals are documented in ASTM standard (ASTM G215 17) [4] and presented in Figure 2.1 [5] with respective reactions. At equilibrium potential, no subsequent loss of metal ions into electrolyte is permitted. However, interestingly metal electrode, when immersed in the electrolyte, does not attain the equilibrium potential rather tends to achieve a more positive side. This is plausible because more electrons are taking part (consumed) during the cathodic reaction. Besides this, many factors contribute to electrochemical cell formation and are explained in the below subsections.

2.2.1 CHARGE TRANSFER AND MASS TRANSPORT

As stated earlier, during electrochemical cell formation, transfer of electrons occurs at the metal electrode and electrolyte interface, which results in charge transfer and mass transport. For typical mass transport phenomena, diffusion, migration, and convection are the three main driving mechanisms for the respective transport of reactants and products from the bulk electrolyte to the metal electrode surface and from the electrode to the electrolyte solution, as shown schematically in Figure 2.2. However, exchange of current density, i_e, and activation barrier, α_0, are responsible for the transfer of charge during electrochemical cell (Figure 2.2). Note that the migration and convection become negligible in the absence of an electric field, and diffusion is the only dominating factor during the process [6].

2.2.2 KINETICS OF ELECTROCHEMICAL CORROSION AND POLARIZATION

It is well-established that if the metal is not corroding, the local potentials are the open circuit potential (OCP), as shown in the schematic in Figure 2.3. The electrode potentials $E_{0(a)}$ and $E_{0(c)}$ are the OCP, while $i_{0(a)}$ and $i_{0(c)}$ represent the exchange current densities. The subscripts "a" and "c" specify anodic and cathodic areas, respectively.

As the metal starts corroding due to the charge transfer and mass transport process, the potential of cathodic areas tends to move towards a more

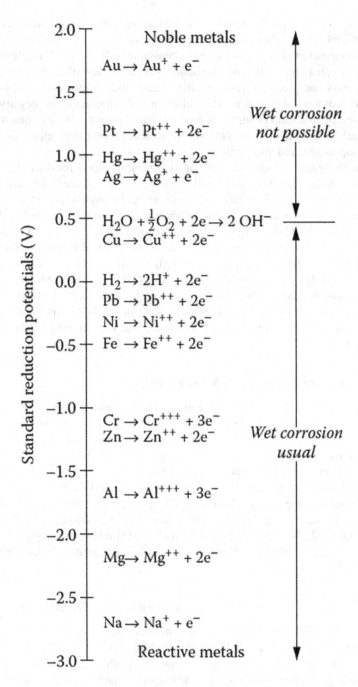

FIGURE 2.1 Standard reduction potential of various metals documented in ASTM G215 17. (Adopted and reprinted with permission from M.F. Ashby, Elsevier, 4th Ed., (2018) pp. 399 [5].)

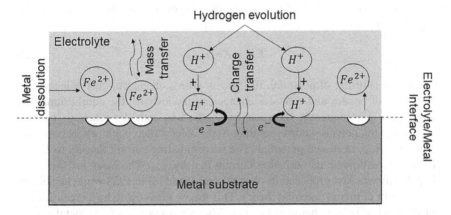

FIGURE 2.2 Schematic representation of charge transfer and mass transport phenomenon during electrochemical reaction in plane carbon steel. (Adopted and modified with permission from Deshpande and Sazou, CRC Press, 1st Ed., (2015) p. 27 [2].)

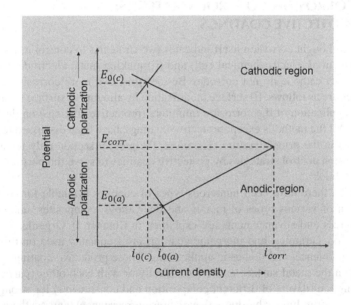

FIGURE 2.3 Schematic representation of kinetics of electrochemical process of pure iron in acid solution. Original concept was proposed by M.G. Fontana, Tata McGraw-Hill, New Delhi, India, 3rd Ed., (2005) [1].

anodic side and vice versa. The subsequent respective increase and decrease in the rate of anodic and cathodic reactions are obvious with further metal dissolution. This process is known as "polarization", where electrode potential shifts from their equilibrium or open circuit values. This polarization sometimes may be referred to as overpotential or overvoltage [7] that can be represented by

$$E_{OP} = E - E_{OCP} \qquad (2.5)$$

During the shifting of potential, the flow of corrosion current gets reduced, and potentials of anodes and cathodes change continuously till the equilibrium state is reached. Consequently, the rates of anodic and cathodic reactions become equal, and the potential corresponding to the steady state is known as corrosion potential, E_{corr}, or mixed potential. The corresponding current density to E_{corr} in Figure 2.3 is known as corrosion current density, i_{corr}. The basis for the above hypothesis was first proposed by Wagner and Traud [8] in 1938 while explaining the concept of mixed potential theory. This theory turns out to be a breakthrough in the field of electrochemical kinetics. The corrosion rate of the metal dipped into an electrolytic solution can be estimated by polarization experiments using Galvanostat/Potentiostat.

2.3 CORROSION CONTROL STRATEGIES: PROTECTIVE COATINGS

The word/thought corrosion itself indicates two strategies to control it: (i) breaking of potential (electrochemical cell) and (ii) making metal electrode cathodic (metals as a cathode do not corrode). Besides this, two more corrosion control strategies are as follows: (i) surface modification by alloying or surface treatments and (ii) application of the corrosion inhibitors (protective coatings) on the metal surface. All the methods except protective coatings, have been discussed in detail elsewhere in the prior literature. Therefore, the present section only focuses on the corrosion control strategies by protective coatings to serve the purpose of the reference book.

To date, there have been numerous types of coatings available for corrosion protection of various types of metals and their alloys, composites, and ceramics (progress and advancements are explained in Chapter 1). Organic, metallic, nonmetallic, ceramics, and inorganic coatings are frequently used on this front for space, defence, and domestic applications. These protective coatings can be applied on the metal surface either individually or with each other (particularly producing multilayers of different composition and constituents) for strong electrical insulation, higher hardness (wear/abrasion resistance), and aesthetic look. The three main driving mechanisms by which coatings protect the metal surface are (i) barrier protection, (ii) inhibition protection, and (iii) sacrificial protection. When the coatings completely isolate the metal surface from the corrosive environment, the so-called barrier protection comes into play. Inhibition protection comes into the picture when coatings release some chemical substances (inhibitors) to slow down the rate of the corrosion process, while sacrificial protection refers to the substance (sacrificial anode) on coatings that corrodes itself and protects the underlying surface (cathodic in corrosion cell) from corrosion.

2.3.1 ORGANIC COATINGS

Owing to their resistance against the flow of moisture, weather, humidity, and ionic species, organic coatings find a unique niche in widespread applications such as painting, powder coatings, plasma polymerization, and sol–gel coatings. As mentioned above, organic coatings serve as a barrier between the underlying metal and corrosive environment by maintaining the durability of the structure. Their efficiency mainly depends on their mechanical properties, adhesion to underlying substrate, and concentration of suspended inhibitors [9]. The typical inhibitors used in organic coatings are metallic, nonmetallic (mostly inorganics), and organic types either in single- or multilayer form. In automobile or defence structures, multilayered organic coatings are often used with a pretreated primer coat of phosphates or chromates.

Organic coatings are classified according to the chemical composition of resins. The typical organic coatings developed so far on this front are vinyl resin, acrylic, alkyd, silicone-modified alkyd, epoxy resin, urethane resin, and polyester resin–based coatings. The pigment is another important substance in organic coatings. It brings specific colour and opacity, protects the resin from UV rays, and offers resistance to water flow. Interestingly, it acts as a volume filler in the total coating, thereby reducing coating production costs. Some of the popular pigments used in organic coatings are zinc oxide, titanium dioxide, zinc sulphite, white lead, and lithopone (a mixture of $BaSO_4$ and $ZnSO_4$). Sometimes to provide the anticorrosive properties to coatings, anticorrosive pigments such as $ZnO/CaSiO_3$ and $MgO/CaSiO_3$ are also used, which then protect the coatings from salts, oxygen, humidity, and corroding fluids/gases such as SO_2 and SO_3 [10]. The effect of pigmentation in organic coatings results in curing and film formation [11,12], improving coefficient of thermal expansion [12], mechanical properties [13], and glass transition temperature, T_g [14].

Different applying techniques of organic coatings are painting, powder coating, plasma polymerization, sol–gel method, and e-coating [9]. Powder coatings are generally applied by electrostatic or thermoplastic powder spraying, while the sol–gel application is multi-stage. Electrophoretic deposition is a two-step process in which charge transfer takes place at metal electrode and solution interface followed by deposition of thin layer compact film. Although these coatings protect the metal from corrosion attack, the factors such as thermal and hygroscopic cyclic environment, UV radiation, and moisture in the surrounding environment led to the physical and chemical ageing of organic coatings [15]. Therefore, there is a continuous demand for alternative coatings to overcome the above limitations.

2.3.2 CORROSION INHIBITORS: METALLIC AND NONMETALLIC INORGANIC COATINGS

One of the limitations of organic coatings is that they do not possess sufficient wear and scratch resistance (hardness), and hence, may not be a suitable candidate where mechanical properties (applications) are of prime importance. In line with this, metallic coatings are the ideal candidates to prevent the underlying metal from corrosion with the synergistic combination of mechanical and electrochemical properties. As such, two types of metallic coatings, (i) cathodic (noble metallic coatings) and (ii) sacrificial anode coatings, can be applied on the metals *via* thermal spray, laser deposition, ion implantation technique, hop dipping and galvanizing, metal cladding, electro/electroless plating, and vapour deposition [16,17]. Most importantly, the thickness of these coatings can be controlled over a range of several *mm* to a few *μm* or *sub-μm*.

The corrosion protection offered by cathodic type coatings is such that the protective coatings are more corrosion resistant than the underlying metals. In sacrificial anode coatings, the top layer of coatings oxidizes (corrodes) faster than the subsequent metal surface (hence they scarify themselves).

Another type of coatings that is of great interest to the coating community is nonmetallic type inorganic coatings due to their excellent corrosion inhibition ability. These coatings, from their classification sense, comprise of hard ceramics, conversion coatings, and anodized coatings [18]. Ceramic coatings are extensively used in various industrial applications to protect metals from an aggressive corroding environment. Ceramic coatings produced *via* the sol–gel method are useful in multiple aerospace applications where dense coatings are required without any defect or pin-holes. They generally offer protection to the underlying metal *via* barrier mechanisms [19,20]. The typical porcelain coatings and glass enamels can be applied on steel sheets and aluminium for improved corrosion protection in various automotive applications. However, one of the limitations of these ceramic coatings is the intrinsic defects and porosity (ceramics are generally porous) despite their excellent mechanical properties.

Different conversion coatings (mostly phosphate and chromate) can be applied on steel and aluminium for automotive applications and decorative purposes [21,22]. These coatings possess excellent electrical insulation, high adherence to the underlying metal and corrosion resistance, and durability and longevity. Phosphate conversion coatings can be deposited on iron, steel, and aluminium *via* chemical, electrochemical, and deep galvanizing treatment [21–24]. Generally, all the conversion coatings are porous, and hence, the addition of inhibitors such as ZnO and TiO_2 in micro and/or nano form (grains) may act as a sealing agent and improve the anticorrosion performance of the coatings [23,24]. Besides this, conversion coatings also enhance the aesthetic look of the underlying metal and are sometimes also used in the metal finishing process. Before 2010, extensive research was conducted on chromate conversion coatings (CCCs). However, due to stringent environmental conditions, many countries have enacted the use of CCCs.

Apart from this, anodized coatings provide a decorative finish to the underlying metal substrate and enhance the bonding ability between primer and metal. Anodizing of metals is generally performed in chromate or sulphuric acid solution. During chromate assisted anodizing *via* electrochemical deposition, metals are immersed into an aqueous solution of chromic acid (CrO_3) at room temperature for 40–60 min at predetermined current levels. However, the same treatment is also applicable for sulphuric acid–mediated anodizing at a temperature not exceeding 20°C. The typical thickness of coatings produced by anodizing ranges from 8 to 30 μm depending upon the applied current and deposition time.

Despite providing good wear resistance and competitive mechanical and chemical properties, the above coatings cannot sense any internal/external damage, corrosion, and micro-cracks and trigger the inhibitors before the damage could happen. Therefore, the coatings community is continuously searching for an alternative that can respond to the external or internal damage and exhibit the stimuli response (i.e., self-healing/repair and self-cleaning). In the below section, advancement in such coatings in light of the self-repairing mechanism is discussed in detail.

2.4   SMART COATINGS FOR CORROSION CONTROL: MECHANISMS AND ADVANCEMENTS

One immediate question to pose at this juncture is: "What are the necessary key factors that we look forward to while protecting any metal from corrosion?" While answering this question corrosion community typically would like to mention two things: (i) barrier coatings and (ii) corrosion inhibitors. Over the years, extensive research performed on developing such coatings is notable. However, the development of these coatings resulted in a fractional improvement of smart stimuli response characteristics such as sensing (which can sense/detect early corrosion) and acting (self-healing/repair). This led to the birth of SCs with multifunctional characteristics that signal/detect the corrosion and repair/heal the damaged or affected area and defects in underlying metals.

SCs provide feedback responses during the corrosion process due to their autoresponsive characteristics. SCs are attractive mainly because of their improved substrate adherence and surface functionality and enhanced optical and electromagnetic properties. In the current scenario, nanoparticles-incorporated polymer SCs have gained significant attention due to their excellent ability of resistance to water penetration (nanocapsules can act as a sealing and healing agent), improved toughness (an indirect measure of strength), and combined effect of permeability and electrical insulation [25,26]. The main corrosion protection mechanism by which SCs function is the self-healing/repair.

Interestingly, the self-healing mechanisms in polymer-based SCs consist of modifying the chemical and physical properties of the polymer itself. For instance, shape memory polymers in SCs can repair internal damage and defects [27]. Furthermore, healing *via* bulk polymers may help to improve the mechanical and thermal properties from a functionality point of view, thereby healing intrinsic or extrinsic defects. In such cases, the temperature can be used as a triggering mechanism. In an attempt to improve the self-healing characteristics of SCs, Liu et al. [28] used atmospheric moisture hybrid polymer (a mixture of organic and inorganic polymer). Further, the observations of Ghosh and Urban [29] have revealed that UV radiation can also be helpful for triggering the inhibitors in oxetane functional and chitosan group incorporated polyurethane coatings. The popular Diels-Alder reaction can also be used for thermo-reversible crosslinks polymer, particularly in polypropylene-based powder coatings, thereby improving the corrosion protection performance [30].

Another type of self-healing/repair mechanism in SCs is swelling. The swelling of the inhibitor takes place by absorbing the water from the external environment, thus healing the defects produced due to various loading/impact conditions. However, in actual practice, a complete closure of defect/damage/crack is not possible [20]. Bentonite clay is often considered as an ideal material for this type of self-healing/repair mechanism in SCs due to its ability to exchange ions during the healing process.

The incorporation of the healing agent into the polymer can also be done *via* capsulations. The incorporated capsule acts as a healing agent carrier that contains inhibitors and signalling agents [31]. The typical polymer healing agents used on this front are alkyds [32,33] and silyl esters [34]. The self-healing mechanism in SCs *via* encapsulation is schematically shown in Figure 2.4. Apart from the macrocapsules, some studies also reported the ability of micro- and nanocapsules as a healing agent for repairing the damage in SCs (Figure 2.4b) [35,36]. Due to external triggering (say mechanical or thermal loading/impact), breaking the capsule is obvious, thereby releasing the healing agent in the affected/damaged area. The systematic studies of Li and Calle [37] reported novel techniques to embed microcapsules in polymer-based SCs for corrosion detection. Their novel approach is based on corrosion detection by developing pH-triggered SCs. However, one of the demerits of encapsulation is the difficulty in judging the volume of the healing agent and its capability to cure/heal the crack or defects. Often with the limited volume of parent material/substrate, it is challenging to incorporate healing agent capsules into SCs. Therefore, there is continuous thrust to develop the capillary-based capsule (also known as "proto-vascular"), which can deliver enough/controlled volume of healing agent.

Apart from organic capsules, other encapsulating/healing agents used on this front are inorganic capsules [38–40], dendrimers [41], polyelectrolytes [42], natural minerals [43,44], and conducting polymers [45,46]. From the above observations, it can be construed that the stability of reactive/healing agents/capsules is necessary throughout the process for durable coating and repair to occur after impact.

2.5 CONCLUSIONS

In this chapter, we brief out the corrosion process and its electrochemical aspects for the general understanding of the readers. Different types of corrosion protection mechanisms are stated and explained in light of protective coatings. Furthermore, other corrosion inhibition coatings are briefly described before proceeding to SCs for corrosion protection applications. The main mechanism responsible for corrosion inhibition *via* SCs is self-healing/repair and it is explained in detail. This chapter aims to get the reader familiar with the fundamentals of the corrosion

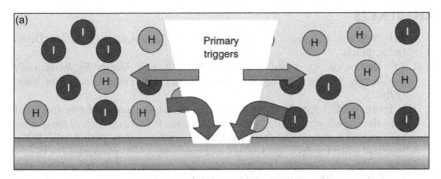

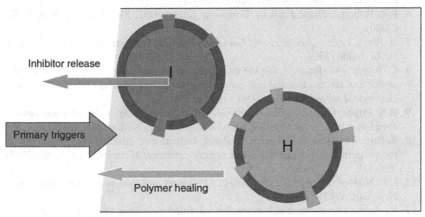

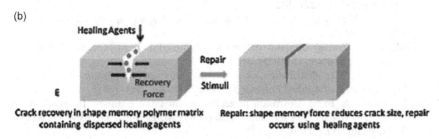

Crack recovery in shape memory polymer matrix containing dispersed healing agents   Repair: shape memory force reduces crack size, repair occurs using healing agents

FIGURE 2.4 Schematic representation of self-healing mechanisms *via* encapsulation in SCs (a) at mesoscale (Adopted and reprinted with permission from A.S.H. Makhlouf, Woodhead Publishing Series in Materials Science and Engineering, Elsevier (2014) [20]) and (b) at micro- and nanoscale (i.e., using micro/nanocapsules of healing agents). (Adopted and reprinted with permission from S.B. Ulaeto et al., *Progress in Organic Coatings*, 111 (2017) 294–314 [36].)

process and its protection *via* different inhibitors. We believe that the readers are now well-versed with the fundamental concepts and underlying corrosion protection mechanisms *via* SCs.

REFERENCES

1. M.G. Fontana, *Corrosion Engineering*, 3rd Ed., Tata McGraw-Hill, New Delhi, India, (2005).
2. P.P. Deshpande, D. Sazou, *Corrosion Protection of Metals by Intrinsically Conducting Polymers*, 1st Ed., CRC Press, Taylor and Francis Group, New York, (2016).
3. V. Fourmond, C. Léger, An introduction to electrochemical methods for the functional analysis of metalloproteins, In: Crichton, R. R., Louro, R. O. (eds.) *Practical Approaches to Biological Inorganic Chemistry*, 2nd Ed. Elsevier, Amsterdam, Netherlands, (2020).
4. Standard guide for electrode potential measurement, ASTM standard G215 17.
5. M.F. Ashby, H. Shercliff, D. Cebon, *Materials Engineering, Science, Processing and Design*, p. 399). Elsevier, Oxford, (2007).
6. P.R. Roberge, *Hand Book of Corrosion Engineering*, McGraw-Hill, New York, (2000).
7. N. Perez, *Electrochemistry and Corrosion Science*, Springer (India) Pvt. Ltd., New Delhi, India, (2004).
8. C. Wagner, W. Traud, Z. On the interpretation of corrosion processes by superimposition of electrochemical cub-processes and on the potential formation at mixed electrode, *Elektrochem*, 44 (1938) 391-454.
9. B.N. Popov, Organic coatings, In: *Corrosion Engineering: Principles and Solved Problems*, 1st Ed., pp. 557–579. Elsevier, Amsterdam, Netherlands, (2015).
10. S. Rosi, M. Fedel, F. Deflorian, S. Zanol, Influence of different color pigments on the properties of powder deposited organic coatings, *Materials Design*, 50 (2013) 332–341.
11. L.E. Nielsen, *Mechanical Properties of Polymers and Composites*, Marcel Dekker, New York, (1974).
12. K. Sato, The mechanical properties of filled polymers, *Progress in Organic Coatings*, 4 (1976) 271–302.
13. A. Zosel, Mechanical behavior of coating films, *Progress in Organic Coatings*, 8 (1980) 47–79.
14. M. Narkis, Crazing in glassy polymers. Polymer-glass bead composites, *Polymer Engineering Science*, 15 (1975) 316–320.
15. M. Narkis, Size distribution of suspension-polymerized unsaturated polyester beads, *Journal of Applied Polymer Science*, 23 (1979) 2043–2048.
16. W. Plieth, A. Bund, Corrosion protection by metallic coatings, In: Bard, A.J., Stratmann, M. (eds.) *Encyclopedia of Electrochemistry*, vol. 4, pp. 567–592. Wiley-VCH, Weinheim, (2003).
17. B.G. Rhee, H.Y. Sohn, Metal alloy coatings: physical, wear-related, and other surface characteristics (London, United Kingdom), *High-Temperature Mater Processes*, 21 (4), pp 217–227, (2002).
18. J.R. Davis, *Corrosion: Understanding the Basics, Corrosion Control by Protective Coatings and Inhibitors*, ASM International, Materials Park, OH, pp. 396–402, (2000).
19. D. Wang, G.P. Bierwagen, Sol–gel coatings on metals for corrosion protection, *Progress in Organic Coatings*, 64(4) (2009) 327–338.
20. A.S.H. Makhlouf, *Handbook of Smart Coatings for Materials Protection*, Woodhead Publishing Series in Metals and Surface Engineering, Elsevier, (2014).
21. N.S. Bagal, V.S. Kathavate, P.P. Deshpande, Nano TiO_2 phosphate conversant coatings- A chemical approach, *Electrochemical Energy Technology*, De Gruyter, 4 (2018) 47–54.

22. V.S. Kathavate, N.S. Bagal, P.P. Deshpande, Corrosion protection performance of nano TiO_2 incorporated phosphate coatings obtained by anodic electrochemical treatment, *Corrosion Reviews*, 37(6) (2019) 565–578.

23. V.S. Kathavate, D.N. Pawar, N.S. Bagal, P.P. Deshpande, Role of nano ZnO particles in the electrodeposition and growth mechanism of phosphate coatings for enhancing the anti-corrosive performance of low carbon steel in 3.5% NaCl aqueous solution, *Journal of Alloys and Compounds*, 823 (2020) 153812.

24. V.S. Kathavate, P.P. Deshpande, Role of nano TiO_2 and nano ZnO particles on enhancing the electrochemical and mechanical properties of electrochemically deposited phosphate coatings, *Surface and Coatings Technology*, 394 (2020) 125902.

25. X. Shi, T.A. Nguyen, Z. Suo, Y. Liu, R. Avci, Effect of nanoparticles on the anti-corrosion and mechanical properties of epoxy coating, *Surface and Coatings Technology*, 204(3) (2009) 237–245.

26. A. Hamdy, Corrosion protection performance via nano-coatings technologies, *Recent Patents Materials Science*, 3(3) (2010) 258–267.

27. D.Y. Wu, S. Meure, D. Solomon, Self-healing polymeric materials: a review of recent developments, *Progress in Polymer Science*, 33(5) (2008) 479–522.

28. H.A. Liu, B.E. Gnade, K.J. Balkus, A delivery system for self-healing inorganic films, *Advanced Functional Materials*, 18(22) (2008) 3620–3629.

29. B. Ghosh, M.W. Urban, Self-repairing oxetane-substituted chitosan polyurethane networks, *Science*, 323 (2009) 1458–1460.

30. H.R. Fischer, Self repairing systems – a dream or reality, *Natural Science*, 2 (2010) 873–901.

31. C. Suryanarayana, K.C. Rao, D. Kumar, Preparation and characterization of microcapsules containing linseed oil and its use in self-healing coatings, *Progress in Organic Coatings*, 63(1) (2008) 72–78.

32. A. Kumar, L.D. Stephenson, J.N. Murray, Self-healing coatings for steel, *Progress in Organic Coatings*, 55(3) (2006) 244–253.

33. S.J. García, H.R. Fischer, P.A. White, J. Mardel, Y. González García, J.M.C. Mol, A.E. Hughes, Self-healing anticorrosive organic coating based on an encapsulated water reactive silyl ester: synthesis and proof of concept, *Progress in Organic Coatings*, 70(2–3) (2011) 142–149.

34. Y. González García, S.J. García, A.E. Hughes, J.M.C. Mol, A combined redox-competition and negative feedback SECM study of self-healing anticorrosive coatings, *Electrochemistry Communications*, 13(10) (2011) 1094–1097.

35. Y. Yang, M.W. Urban, Self-healing polymeric materials, *Chemical Society Reviews*, 42(17) (2013) 7446–7467.

36. S. B. Ulaeto, R. Rajan, J. K. Pancrecious, T.P.D. Rajan, B.C. Pai, Developments in smart anticorrosive coatings with multifunctional characteristics, *Progress in Organic Coatings*, 111 (2017) 294–314.

37. W. Li, L.M. Calle, Micro-encapsulation for corrosion detection and control, *Proceedings of the 1st International Conference on Self-Healing Materials*, The Netherlands, April 2007, Noordwijkaan Zee, 2007, pp. 18–20.

38. S.P.V. Mahajanarn, R.G. Buchheit, Characterization of inhibitor release from Zn-Al-V10O28(6-) hydrotalcite pigments and corrosion protection from hydrotalcite-pigmented epoxy coatings, *Corrosion*, 64(3) (2008) 230–240.

39. G. Williams, H.N. McMurray, Inhibition of filiform corrosion on polymer coated AA2024-T3 by hydrotalcite-like pigments incorporating organic anions, *Electrochemical and Solid State Letters*, 7(5) (2004) B13–B15.

40. E. Abdullayev, Y. Lvov, Clay nanotubes for corrosion inhibitor encapsulation: release control with end stoppers, *Journal of Materials Chemistry*, 20(32) (2010) 6681–6687.
41. A. Trinchi, T.H. Muster, A review of surface functionalized amine terminated dendrimers for application in biological and molecular sensing, *Supramolecular Chemistry*, 19(7) (2007) 431–445.
42. D.V. Andreeva, E.V. Skorb, D.G. Shchukin, Layer-by-layer polyelectrolyte/inhibitor nanostructures for metal corrosion protection, *ACS Applied Materials and Interfaces*, 2(7) (2010) 1954–1962.
43. J. Sinko, Challenges of chromate inhibitor pigments replacement in organic coatings, *Progress in Organic Coatings*, 42(3–4) (2001) 267–282.
44. D. Raps, T. Hack, J. Wehr, M.L. Zheludkevich, A.C. Bastos, M.G.S. Ferreira, O. Nuyken, Electrochemical study of inhibitor-containing organicinorganic hybrid coatings on AA2024, *Corrosion Science*, 51(5) (2009) 1012–1021.
45. M. Kendig, M. Hon, L. Warren, Smart corrosion inhibiting coatings, *Progress in Organic Coatings*, 47(3–4) (2003) 183–189.
46. D. Kowalski, A. Tighineanu, P. Schmuki, Polymer nanowires or nanopores? Site selective filling of titania nanotubes with polypyrrole, *Journal of Materials Chemistry*, 21(44) (2011) 17909–17915.

3 Smart Protective Coatings

Automobile, Aerospace, and Defence Applications

3.1 INTRODUCTION

Human life in the 21st century continuously evolves around the development of functional and engineering materials. The perfect reason to believe this argument is the increased use of novel materials in various real-life/domestic applications. These domestic and classified applications may range from as small as micro/nano-electro-mechanical systems (i.e., MEMS and NEMS devices) at μm and nm scale to as large as automobile parts, space and defence structures at mesoscale. Over the years, development in these materials/structures for improved performance and longevity has been notable. However, this development in functional and engineered materials is reflected in the fractional improvement in their corrosion resistance (or electrochemical properties) though their functional properties (for example, mechanical or electrical or magnetic) are greatly improved. Most of the materials in their applications are subjected to the dynamic environment under coupled loading conditions (say electro-mechanical or thermo-mechanical), which eventually causes the deterioration in their electrochemical properties. For example, aluminium alloys (in the form of thermal blankets) used in satellites and space telescopes have to withstand external environmental impact and reflect the solar heat radiations (i.e., thermo-mechanical environment) to protect the structures from heat/thermal shock. While the aluminium (or alloy) surface in the above application is exposed to an open environment, its protection from the corrosive agents is of paramount importance. Many advanced and conventional materials demand well-defined surface and electrochemical properties for valorizing the material for a specific application. Therefore, their surface functionalization has become extremely important from the enhanced functionality point of view.

The term "surface functionalization" refers to a channel to modify/tune or enhance the surface properties, thus suiting material for a specific application. A modified surface may comprise new and/or modified group of characteristics such as physical and chemical properties; enhanced mechanical, optical, and electrical properties; and tuned morphological features. The schematic presented in Figure 3.1 systematically demonstrates the overview of functional characteristics

DOI: 10.1201/9781003200635-3

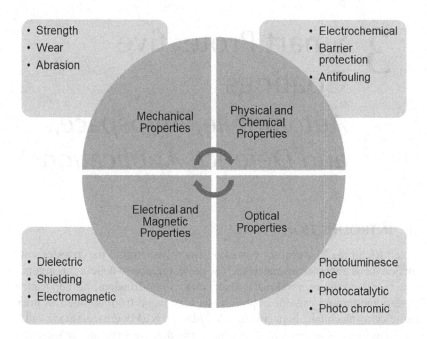

FIGURE 3.1 Schematic representing the overview of functional properties tuned *via* surface functionalization.

that can be tuned or modified *via* surface functionalization of the materials used in automotive, defence, and aerospace structures.

Of the several surface functionalization techniques, the application of smart coatings (SCs) is perhaps the most convenient route to protect the surface and enhance its physicochemical properties as listed in Figure 3.1. One good example of such application is applying the epoxy polymer-based SCs encapsulated with cerium and polysiloxanes healing agent to protect steel and aluminium in automotive parts [1,2]. In the below sections, such applications of SCs for automobile, aerospace, and defence structures are briefly discussed with the aid of recent innovations in both materials and coatings.

3.2 ADVANCES IN CONVENTIONAL MATERIALS AND SURFACE PRETREATMENTS

As discussed in the previous chapters, materials are the greatest gift to human life. These materials (metals or their alloys) design has a history of the middle palaeolithic (stone) age (~50,000 BC). However, the continuous revolution of metallic materials for the better sustainability of human life is notable.

Owing to their superior mechanical properties, iron (and its alloys) is the principal structural material. It is reported that the first use of iron is thought to be approximately 4000 years ago where humans used meteoric iron (an alloy containing nickel and iron) [3,4]. After iron, aluminium (and its alloys) is the most widely preferred material for structural applications due to its high ductility, strength to weight ratio, and enhanced corrosion resistance (compared to iron). Due to its lightweight characteristics, aluminium is recognized as one of the "space-qualified materials" widely used by many aerospace and defence industries. Although it is among abundantly available materials in Earth's crust after oxygen and silicon, extracting aluminium from its ore alum is a challenging task, which subsequently increases its production costs. The materials scientists have also developed a composite of an aluminium matrix with some reinforced fibres (also known as Al-metal matrix composites, MMC) to meet the growing demand of automobile, aerospace, and defence industries, which further resulted in improved thermal, mechanical, and electrical properties. However, one of the downsides of Al-MMC is the structural inhomogeneity, which reflects in the preferential dissolution of particles at the matrix/fibre interface and subsequent deterioration in corrosion resistance.

In the current scenario, the weight reduction of automotive, aerospace, and defence structures is of paramount importance to improve fuel consumption from a transportation point of view. In line with this, magnesium, which has a density of ~25% steel and ~60% aluminium, is ideal for the transportation industry. Some newly developed magnesium alloys (i.e., Elektron, AZ31HP-O, and AZ91E), particularly developed with doping the rare earth elements, can withstand against impact loading in various electronics, automobile, and aerospace applications. However, their poor corrosion resistance is still a major concern for the materials community. The main focus of the present chapter is to emphasize the applications of SCs on structural materials that are being used for automobile, aerospace, and defence structures. Therefore, we limit our discussion to these common but important structural materials, thereby serving the purpose of the chapter.

As mentioned above, surface functionalization of these materials is an important aspect from enhanced functionality and longevity point of view. Keeping this in view, numerous deposition strategies have been adopted/invented by well-recognized research groups worldwide. In several deposition techniques, surface pretreatment becomes extremely important from the improved thickness, adherence, and aesthetic point of view for protecting corrosion in automobile and aerospace materials. The simplest way of surface pretreatment is applying conversion coatings on the metal surface. Initially, chromate conversion coatings (CCCs) were developed and applied as a primer that further reflected in high adherence to the underlying metal and top coat, lucrative surface appearance, and enhanced surface functionalization. However, many countries have enacted the use of toxic hexavalent chromate as it affects the human nervous system and creates difficulties in breathing. Therefore, numerous chromate-free conversion coatings have recently been developed to replace poisonous chromate with enhanced surface protection. For instance, recent work by Kathavate and co-workers [5–8] demonstrated the ability of phosphate conversion coatings with the improved corrosion resistance of steel used in automobile structures. The deposition of metal dioxide–incorporated phosphate layer on low carbon steel subsequently enhanced the corrosion resistance by ~8 times (comparative to bare steel). The notable works cited in the references [9–13] also reported the improved surface characteristics of phosphate-coated Mg alloys for automobile applications and transportation industries. The extensive work by Hamdy and co-workers [14–18] illustrates the variety of chromate-free conversion coatings applied on different automobile and aerospace materials/structures. All these CCCs cited above have been deposited by incorporating nano/micro metal dioxide particles, thus providing better crystalline structure, increased surface coverage, resistance against the penetration of electrolyte, and an impressive aesthetic look.

To sum up, the research and development of the surface pretreatment process for the protection of automobile and aerospace materials is a vast area. From the future perspectives, the main driving force for the innovations and development in surface pretreatment process (or CCCs) is (i) growing demand of automobile and aerospace industries to produce the CCCs with high functionality relatively at lower costs and (ii) eliminating/replacing the toxic elements such as chromate and volatile organic coatings (VOCs) from the coating process. In view of the above aspects, we discuss the advancement in SCs for the applications in corrosion protection of automobile and aerospace materials.

3.3 ADVANCES IN PROTECTIVE COATINGS FOR AUTOMOBILE, AEROSPACE, AND DEFENCE APPLICATIONS

As discussed in the previous chapters, SCs can be classified according to their deposition/fabrication routes, applications, responsiveness, inhibitor materials used, triggering mechanisms, and function. The main advantage of SCs in various applications is the multifunctionality associated with them. Over the years, considerable progress has been made to improve the multifunctionality of SCs for the corrosion protection of automobile, aerospace, and defence materials/structures. Therefore, to rationalize the further discussion, we mention the selected SCs based on their functionality (for example, self-cleaning, corrosion sensing, and pressure sensing), which has a diverse range of applications in the above sectors.

3.3.1 WATER-SOLUBLE COATINGS/PAINTS (WSPs)

In the past two decades, considerable research in water-soluble coatings/paints (WSPs) led to strong competition among the coatings industry. These WSPs are cheaper and have a very good appearance, making them useful for several automobile applications. Furthermore, compared to alkyl-based and other paints, WSPs produce/emit fewer toxic elements and lesser flashy odour. Most importantly, they eliminate the possibility of making harmful/toxic VOC and possess good chemical stability. In the present scenario, WSPs find a unique niche in several automobile applications such as painting the interior of the car and bike, chassis, and decoration purposes. For several years or so on, WSP will remain a strong competitor among the paint and coatings industries due to the advanced developments in resins [19].

3.3.2 Ultraviolet Curable Coatings (UVCCs)

Another important class of coatings after WSP is ultraviolet curable coatings (UVCCs), which are VOC and hazardous air pollutant (HAP) free. Unlike WSP and other coatings, UVCC cures faster (probably within a second) and reflects higher throughput and production costs reduction. Therefore, there is a growing demand for fast curing and low emitting VOC and HAP particles for automobile applications. The main principle behind UVCC is the transformation of coating liquid into cross-linked solid polymer *via* photochemical process (UV curing). This also eliminates the need for isocyanate cross-linking agents. The classification of UVCC is based on free radical type (acrylate) and cationic (mostly epoxy). The applications of UVCC reflect energy savings by a huge margin and the elimination of organic solvent.

These UVCCs can be combined with CCCs and non-CCCs. The complexities associated with the coating system can be eliminated by combining the top coat and base primer into a single layer. However, incorporation of CCCs in UVCC would result in the emission of toxic hexavalent chromium. Therefore, it is highly recommended to use the non-CCCs with UVCC for lowering the emission of poisonous products with improved efficiency. In line with this, currently, efforts are directed to combine the topcoat layer (organic) and underlying primer (inorganic) into a single layer to reduce the production cost (as shown in Figure 3.2) [20–22]. This mixed UVCC–non-CCCs paint finds applications in various automobile and aerospace parts, wherein these paints can be applied *via* spraying, brushing, and spinning in a single step.

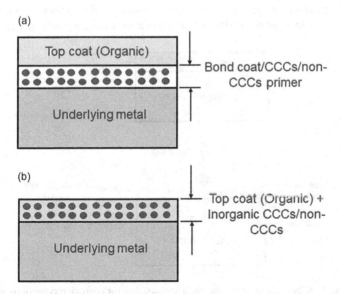

FIGURE 3.2 Schematic representing the (a) conventional three multilayer coatings system and (b) UVCC system (by mixing top organic and inorganic primers).

3.3.3 Multifunctional SCs: Mechanisms and Recent Progress

In the current scenario, the heavy competition between automobile and aerospace industries has led the materials scientists to design and develop the coatings that can heal/repair, sense, and trigger themselves upon mechanical and/or chemical (or external environment) degradation. The development of SCs fulfils these needs by responding actively after sensing the changes in the surrounding environment while remaining passive for their functionality. The development of SCs for the protection of automobile and aerospace materials/structures is based on the following viewpoints: (i) it should heal the cracks or scratches occurring inside the coatings (internal changes) and (ii) it should also respond to the changes in the surrounding environment such as pH, salinity, temperature, and pressure (external environmental changes). While the former represents the stimuli response mechanisms, the latter indicates the self-sensing characteristics.

On this front, active corrosion sensing and pressure sensing SCs (hereafter referred to as CSSCs and PSSCs, respectively) are preferred in many instances and are of paramount importance. While the CSSCs are used to prevent the corrosion events during early stages of corrosion (by sensing/detecting) occurring in automobile and aerospace parts due to aggressive environment, PSSCs have been preferred in space vehicles and wind tunnel models to sense the changes in surrounding pressure, thus suiting the acoustics and aerodynamics related applications.

These CSSCs are mostly pH-sensitive and change the colour of the surface in response to the change in external surroundings (i.e., corrosive media). They contain colour changing dyes incorporated within a matrix, which further alters the colour upon the formation of corrosion products on the surface and/or change in pH of the site (due to oxidation), as shown by the schematic in Figure 3.3.

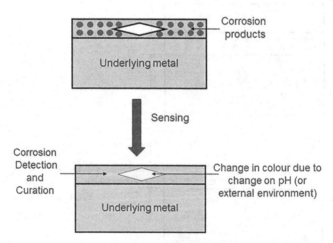

FIGURE 3.3 Schematic representation of self-sensing (or corrosion detection) SCs. The colour of the dye changes after the early detection of corrosion events (or corrosion products). For colour proofs, readers are advised to refer to the online copy of the article.

In many instances, transparent matrices are used to visualize the colour changes in the dyes upon the corrosion attack. The colour changing dyes are made with anticorrosive agents and microcapsules (as additives). Interestingly, some dyes also contain incorporated pigments, which then do not change the colour but release the anticorrosive agents/inhibitors upon localized damage or corrosion attack. In view of this, the coating formulation and developments in paints have gained significant attention due to recent advancements in anticorrosive pigments and dyes. For instance, anticorrosive agents such as hydroxyquinolines, fluorescein, oxines, bromothymol blue, and 7-amino-4-methylcoumarin, 7-diethylamino-4-methylcoumarin have been extensively used in CSSCs for the protection of aerospace materials/structures [23].

Among all CSSCs, the pH-sensitive SCs are quite popular due to their ability to sense the change in pH of the surrounding environment. For example, Li and Calle [24] invented the novel microencapsulated pH sensing coatings for corrosion detection applications. The sensing ability of microcapsules was increased by increasing the content of cross-linking species and a simultaneous decrease in wall thickness. For corrosion detection, the change in colour of microcapsules was observed in basic pH solution. In another interesting study by Maia et al. [25], the corrosion sensing ability of pH-sensitive SCs formulated with 12 wt% polyureas as microcapsule and phenolphthalein as corrosion sensing agent was successfully demonstrated on AA2024 and AZ31 alloys, thus making them suitable for automobile applications. The SCs were developed to sense the corrosion and heal/repair the damage in aerospace structures, thereby reducing the routine maintenance cost and eliminating the time-consuming repair works due to corrosion damage [26]. The micro/nanocapsules used in these pH-sensitive SCs are passive in their mode of action [23,26–30].

Li et al. [31] have successfully demonstrated the corrosion detection in the aluminium alloy by producing the SCs made of phenylfluorone incorporated in acrylic paint. The corroded area on the underlying metal was identified by detecting the fluorescence quenching in SCs. Note that the thickness of CSSCs is an important consideration along with the fluorescence or colour changing ability of dyes for the above application. However, recently Loh and co-workers [32] reported the ability of carbon nanotube polyelectrolyte composite multilayer thin film to detect changes in electrochemical properties and mechanical strain. Although the above studies [26–32] had demonstrated the ability of pH-sensitive SCs for the corrosion protection of aeronautical and automobile structures, the cost-effective production of these CSSCs is still a timely and open subject for further research.

Besides corrosion detection, the active sensing SCs also find a unique role in surface pressure sensing applications such as aeronautical vehicles and wind tunnel models. This has provided a breakthrough in aerodynamics and acoustics (both underwater and air). These PSSCs sense the change in surface pressure during the operation *via* the change in the intensity of emitted light. The basic operational principle behind the PSSCs is that the metal surface is painted by luminescent paint, which will be excited by the light of a specific wavelength.

Following this, any changes in the surface pressure will be sensed and indicated as a change in colour, which is further captured by the digital camera. While measuring the surface pressure in aerodynamics and acoustics devices *via* PSSCs, the response time of these coatings is an important aspect. It is reported in the literature that the ideal response time of PSSCs to sense/detect any changes in local surface pressure should be less than 1 μs [33]. Figure 3.4 illustrates the working principle of PSSCs owing to their faster response time (i.e., less than 1 μs or fraction of seconds).

The response time constraint factor has led to heavy competition among coating industries to make efficient fast-acting (sensing) SCs cost effective. In line with this, a wide variety of PSSCs is reported in the literature for various aeronautical and acoustics applications. For instance, Gregory et al. [34] have successfully tested the porous PSSCs for fluid flow and acoustics applications. The porous PSSCs can hold the binder (porous materials) and offer large surface coverage. Their response time is usually less than 1 μs, making them suitable for pressure sensing applications in high-pressure environments (for example, underwater applications where the material is subjected to huge pressure ~600 bar). The typical porous materials used in porous PSSCs are anodized titanium and aluminium, polymer, and ceramics [34]. One of the important features of porous PSSCs is the fast formulation associated with the porous binders, which further offers the rapid diffusion of oxygen and surface interactions. This enables the PSSCs to respond quicker and repair the surface [34].

Ruthenium (Ru)-incorporated tris(4,7-diphenylphenanthroline) PSSC has been produced on anodized aluminium by Kameda and co-workers [35] to measure the nonperiodic distribution of surface pressure on delta wing subjected to high angle attack of transonic flow. The pressure distribution in the transonic environment was measured by imaging the coating colour change due to interaction at surfaces between the leading edge and shock waves with the ultra-high resolution camera.

In the past decade, graphene has been emerged as one of the promising materials for energy harvesting, with various mechanical and electronics applications. Owing to their superior mechanical properties (elastic modulus, E ~1.4 TPa and

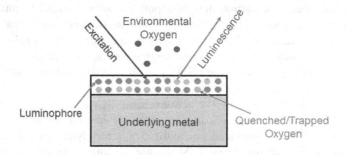

FIGURE 3.4 Schematic representation of working principle of PSSCs. (Adopted and modified with permission from S.B. Ulaeto et al., *Progress in Organic Coatings*, 111 (2017) 294–314 [23].)

high stiffness), graphene in coatings has attracted the entire coatings community, and many of its characteristic features are still unexploited. One such feature of graphene is as a sensor in SCs or PSSCs. Few studies [36,37] have reported the application of functional graphene sensors, particularly Raman-strain sensors, to measure strain by sensing the change in the external environment. The single-crystal graphene flakes were mechanically exfoliated and incorporated in polymethyl methacrylate to formulate the graphene sensor-based PSSCs, which were more sensitive to the strain than other PSSCs. On the other hand, chemically vapour deposited (CVD) graphene thin film is also expected to offer high accuracy ~0.01% and precise resolution ~27–30 $\mu\varepsilon$. In another attempt, Gong et al. [37] had produced the thin PSSCs by sandwiching gold nanowire tissue paper between two thin polydimethylsiloxane sheets for optoelectronics applications. The produced PSSCs offered high wear resistance and mechanical flexibility (even at low pressure and fast response time).

The extent of these PSSCs could also find a wide range of applications in real-time monitoring of blood pulses (blood pressure apparatus). Table 3.1 summarizes CSSCs and PSSCs that have been extensively researched at laboratory scale so far, along with their functionality, type of micro/nanoencapsulation, and applications. Although various PSSCs have been invented/developed to date (authors have extensively reviewed research and review articles till June 2020), the commercial applications of these coatings are still far from meeting industrial acceptance. One of the important challenges for applying PSSCs at a commercial scale is the optimization in coating parameters (for specific applications). In view of this, some of the important aspects regarding the optimization in coatings parameters are discussed below. These parameters are of paramount importance from design and development perspectives for a specific application.

3.4 OPTIMIZATION IN PROCESS PARAMETERS AND RELATED ISSUES

The moment we call ourselves an engineer, the first thought that should come to our mind is to deliver/produce the best quality of results/products by effectively using available resources in a cost-effective manner. At the same time, quality of the product should not be compromised. The huge competition in SCs led scientists to optimize the processing parameters for the development of durable and high throughput coatings simultaneously with low production costs. Several process parameters contribute to the multifunctional characteristics of protective SCs, and their optimization is of paramount importance.

While there is a growing demand for CSSCs in the automobile and aerospace industries for anticorrosive applications, the optimization of sensing agents is of primary importance. On this front, one of the fundamental approaches for active sensing coatings is the effective use of dyes and nanocontainers (capsules) capable of sensing the pre-corrosion events and subsequently repairing the pre-corroded surface. However, on the other hand, micro/nanocapsules may increase

TABLE 3.1

Summary of CSSCs and PSSCs for Automobile and Aerospace Applications [38]

Coating Type	Functionality	Encapsulation	Industrial Applications
CSSCs	Sensing, colour brilliance, shading, and dispersion	Micro/nano oxides such as TiO_2, ZnO, SiO_2, Fe_2O_3, Fe_2O_4, and Cr_2O_3	Automotive
CSSCs	Corrosion detection	SCs made by incorporating micro/nano clay, TiO_2, and ZnO on aluminium, steel, zinc, and magnesium alloys	Automotive and aerospace
CSSCs	Active corrosion sensing and curation (of pre-corroded surface)	Polymers (particularly gels), hybrid polymers (organic–inorganic), synthetic amorphous silica, SiO_2, and Al_2O_3	Automotive and aerospace
PSSCs	Surface pressure sensing and flow characteristics	Synthetic amorphous silica	Aerodynamics and acoustics

Note: The functionality and type of micro/nanoencapsules used while formulating these coatings is also mentioned for better understanding.

the production cost of such coatings. Therefore, there is wide scope for optimizing micro/nanocontainers and sensing agents (dyes) in the upcoming decade. In line with this, these active sensing mechanisms could also be connected with the sustainability and durability of CSSCs. Besides this, leading approaches to produce the multifunctional CSSCs should also involve the presence of agents for creating the superhydrophobic surface required for water-repelling applications, particularly in automobile industries (i.e., transport sectors). Some of the important process parameters to optimize on this front are pH and ion indicators, which further respond to the environmental triggers.

Another important characteristic of SCs in automobile (or transportation) structures is water repellent tendency. Keeping this in view, maintaining the superhydrophobic surface with low surface roughness (possibly at *nm* level) and surface energy is of great concern. Therefore, there is much scope among coatings industries for the optimized formulation of SCs (i.e., proportionate mixing of soft/smooth polymer and sensing/healing/repair agent) for obtaining ultra-smooth surface relatively at lower production costs. To encounter such issues, the effective use of eco-friendly biocides may be helpful to avoid the agglomeration of antifouling species on the surface (which subsequently degrade the surface smoothness).

Above all these parameters, the deposition technique also plays a vital role in the performance of the coatings designed for a specific application. Therefore, optimizing various deposition/production techniques will eventually lower the production cost and efforts. For instance, Li et al. [39] have optimized various triggering and releasing parameters/agents such as pH, dyes, encapsulation of nanofibres, and releasing time to produce the efficient dual pH $Ce(NO_3)_3$ -incorporated chitosan/polyacrylic acid polyelectrolyte coacervate CSSCs on AA2024-T3 alloys. The produced SCs have offered better anticorrosion protection with enhanced response time, which provided a breakthrough in aerospace industries.

Apart from these experimental parameters, in the upcoming decade, numerical and analytical tools for optimizing process parameters in SCs can also be exploited on large scales. On this front, the artificial neural network (ANN) has gained significant attention for predicting the data sets and optimized values of the input process and functional output parameters. The artificial intelligence/machine learning approach will be an added advantage for designing the SCs with improved functionalities. One of the important advantages of these optimization tools is the minimal use of human efforts and physical resources, which further eliminates trial and error (conventional) policy. There is great scope to exploit the predictive capabilities of the above numerical methods as the obtained data sets help in deciding the input-controlled process parameters, which cannot be controlled during the production/synthesis process. Thus far, such aspects related to SCs are largely unexplored.

3.5 CONCLUSIONS

Owing to their achievable excellent multifunctionality, the research on protective SCs is not exhaustive by any means. The present competition between automobile and aerospace industries has undoubtedly led the coatings scientist to modify the current synthesis, applying and optimization techniques, and accept the new ones. Furthermore, producing these SCs for automobile and aerospace materials is a challenging task, and it indeed needs a suitable engineering/deposition technique. However, while adopting this, coatings industries have to ensure the yield of quality SCs. For instance, using nanocontainers (or nanocapsules) while formulating active sensing SCs would be useful and may help to tune the multifunctional characteristics. However, the discussion presented above leads to the following important conclusions.

The definition of SCs needs to be modified on the basis of optimization and standard testing techniques with the aid of economical and environmental viewpoints. This will be also helpful in the classification of SCs according to their multifunctional characteristics. Several test standards have been developed so far for optimizing the coating parameters. However, the future of protective SCs for automobile, aerospace, and military applications is based on the following: (i) development of protective SCs for protecting the lightweight alloys such as magnesium alloys and composite materials; (ii) advancement in polymer chemistry and applying techniques; (iii) development of SCs for "smart maintenance system" in automobile and aerospace structures, which include a combination of safety measures and cost reduction; and (iv) development of advanced and efficient pre-surface treatment process, which will expedite the response time in SCs. Most importantly, producing durable SCs is a great concern and added multifunctionality increases the production/fabrication cost. Therefore, there is a tremendous scope to overcome these aspects in the next decade and so on.

REFERENCES

1. S.H. Cho, S.R. White, P.V. Braun, Self-healing polymers: self-healing polymer coatings, *Advanced Materials*, 21(6) (2009) 645–649.
2. D. Snihirova, S.V. Lamaka, M.F. Montemor, "SMART" protective ability of water-based epoxy coatings loaded with $CaCO_3$ microbeads impregnated with corrosion inhibitors AA2024 substrates, *Electrochimica Acta*, 83 (2012) 439–447.
3. T.A. Rickard, The use of meteoric iron, J. Royal Anthropological Institute of Great Britain and Ireland (Royal Anthropological Institute of Great Britain and Ireland) 71(1/2) (1941) 55–66.
4. A.S.H. Makhlouf, *Handbook of Smart Coatings for Materials Protection*, Woodhead Publishing Series in Metals and Surface Engineering, Elsevier, Cambridge, (2014).
5. N.S. Bagal, V.S. Kathavate, P.P. Deshpande, Nano TiO_2 phosphate conversant coatings – a chemical approach, *Electrochemical Energy Technology*, De Gruyter, 4 (2018) 47–54.
6. V.S. Kathavate, N.S. Bagal, P.P. Deshpande, Corrosion protection performance of nano TiO_2 incorporated phosphate coatings obtained by anodic electrochemical treatment, *Corrosion Reviews*, 37(6) (2019) 565–578.
7. V.S. Kathavate, D.N. Pawar, N.S. Bagal, P.P. Deshpande, Role of nano ZnO particles in the electrodeposition and growth mechanism of phosphate coatings for enhancing the anti-corrosive performance of low carbon steel in 3.5% NaCl aqueous solution, *Journal of Alloys and Compounds*, 823 (2020) 153812.
8. V.S. Kathavate, P.P. Deshpande, Role of nano TiO_2 and nano ZnO particles on enhancing the electrochemical and mechanical properties of electrochemically deposited phosphate coatings, *Surface and Coatings Technology*, 394 (2020) 125902.
9. X.B. Chen, X. Zhou, T.B. Abbott, M.A. Easton, N. Birbilis, Double-layered manganese phosphate conversion coating on magnesium alloy AZ91D: insights into coatings formation, growth and corrosion resistance, *Surface and Coatings Technology*, 217 (2013) 147–155.
10. N. Razaee, M.M. Attar, B. Ramezanzadeh, Studying corrosion performance, microstructure and adhesion properties of a room temperature zinc phosphate conversion coating containing Mn^{2+} on mild steel, *Surface and Coatings Technology*, 236 (2013) 361–367.
11. H. Zhang, G. Yao, S. Wang, Y. Liu, H. Luo, A chrome-free conversion coating for magnesium-lithium alloy by a phosphate-permanganate solution, *Surface and Coatings Technology*, 202 (2008) 1825–1830.
12. C. Zhang, B. Liu, B. Yu, Z. Lu, Y. Wei, T. Zhang, J.M.C. Mol, F. Wang, Influence of surface pretreatment on phosphate conversion coating on AZ91 Mg alloy, *Surface and Coatings Technology*, 359 (2019) 414–425.
13. Z. Rajabalizadeh, D. Seifzadeh, Strontium phosphate conversion coating as an economical and environmentally-friendly pretreatment for electroless plating on AM60B magnesium alloy, *Surface and Coatings Technology*, 304 (2016) 450–458.
14. A.S. Hamdy, M. Farahat, Chrome-free zirconia-based protective coatings for magnesium alloys, *Surface and Coatings Technology*, 204 (2010) 2834–2840.

15. A.S. Hamdy, I. Doench, H. Möhwald, Assessment of a one-step intelligent self-healing vanadia protective coatings for magnesium alloys in corrosive media, *Electrochimica Acta*, 56 (2011) 2493–2502.
16. A.S. Hamdy, I. Doench, H. Möhwald, The effect of vanadia surface treatment on the corrosion inhibition characteristics of advanced magnesium Elektron 21 alloy in chloride media, *International Journal of Electrochemical Science*, 7 (2012) 7751–7761.
17. A.S. Hamdy, F. Alfosail, Z. Gasem, Eco-friendly, cost-effective silica-based protective coating for an A6092/SiC/17.5p aluminum metal matrix composite, *Electrochimica Acta*, 89 (2013) 749–755.
18. A.S. Hamdy, D. Butt, Novel smart stannate based coatings of self-healing functionality for magnesium alloys, *Electrochimica Acta*, 97 (2013) 296–303.
19. S.B. Ulaeto, J.K. Pancrecious, T.P.D. Rajan, B.C. Pai, Smart coatings, In: *Noble Metal-Metal Oxide Hybrid Nanoparticles*, pp. 341–372. Elsevier, (2019). https://doi.org/10.1016/B978-0-12-814134-2.00017-6.
20. C. Seubert, K. Nietering, M. Nichols, R. Wykoff, S. Bollin, An overview of the scratch resistance of automotive coatings: exterior clearcoats and polycarbonate hardcoats, *Coatings*, 2(4) (2012) 221–234.
21. R.H. Fernando, Chapter 1: Nanocomposites and nano structured coatings: recent advancements, In: Fernando, R., Sung, L. (eds.) *Nanotechnology Applications in Coatings*, vol. 1008 pp. 2–22. ACS Symposium Series, (2009).
22. J. Mathew, J. Joy, S.C. George, Potential applications of nanotechnology in transportation: a review, *Journal of King Saud University-Science*, 31(4) (2019) 586–594.
23. S.B. Ulaeto, R. Rajan, J.K. Pancrecious, T.P.D. Rajan, B.C. Pai, Developments in smart anticorrosive coatings with multifunctional characteristics, *Progress in Organic Coatings*, 111 (2017) 294–314.
24. W. Li, L.M. Calle, Micro-encapsulation for corrosion detection and control, *Proceedings of the 1st International Conference on Self-Healing Materials*, The Netherlands, April 2007, Noordwijkaan Zee, 2007, pp. 18–20.
25. F. Maia, J. Tedim, A.C. Bastos, M.G. Ferreira, M.L. Zheludkevich, Active sensing coating for early detection of corrosion processes, *RSC Advances*, 4(34) (2014) 17780–17786.
26. L.M. Calle, P.E. Hintze, W. Li, J.W. Buhrow, Smart coatings for autonomous corrosion detection and control, *AIAA SPACE Conference & Exposition* (2010) 8877.
27. W. Feng, S.H. Patel, M.Y. Young, J.L. Zunino, M. Xanthos, Smart polymeric coatings-recent advances, *Advances in Polymer Technology*, 26(1) (2007) 1–13.
28. A. Stankiewicz, I. Szczygiel, B. Szczygiel, Self-healing coatings in anti-corrosion applications, *Journal of Materials Science*, 48(23) (2013) 8041–8051.
29. Y. Yang, M.W. Urban, Self-healing polymeric materials, *Chemical Society Reviews*, 42(17) (2013) 7446–7467.
30. J. Tedim, S.K. Poznyak, A. Kuznetsova, D. Raps, T. Hack, M.L. Zheludkevich, M.G.S. Ferreira, Enhancement of active corrosion protection via combination of inhibitor-loaded nanocontainers, *ACS Applied Materials & Interfaces*, 2(5) (2010) 1528–1535.
31. S.M. Li, H.R. Zhang, J.H. Liu, Preparation and performance of fluorescent sensing coating for monitoring corrosion of Al alloy 2024, *Transactions of Nonferrous Metals Society of China (English Edition)*, 16 (2006) 3–8.
32. K.J. Loh, J. Kim, J.P. Lynch, N.W.S. Kam, N.A. Kotov, Multifunctional layer-by-layer carbon nanotube polyelectrolyte thin films for strain and corrosion sensing, *Smart Materials and Structures*, 16(2) (2007) 429–438.

33. J.H. Bell, E.T. Schairer, L.A. Hand, R.D. Mehta, Surface pressure measurements using luminescent coatings, *Annual Review of Fluid Mechanics*, 33 (2001) 155–206.

34. J.W. Gregory, H. Sakaue, T. Liu, J.P. Sullivan, Fast pressure-sensitive paint for flow and acoustic diagnostics, *Annual Review of Fluid Mechanics*, 46(1) (2014) 303–330.

35. M. Kameda, T. Tabei, K. Nakakita, H. Sakaue, K. Asai, Image measurements of unsteady pressure fluctuation by a pressure-sensitive coating on porous anodized aluminium, *Measurement Science and Technology*, 16(12) (2005) 2517–2524.

36. A.P.A. Raju, A. Lewis, B. Derby, R.J. Young, I.A. Kinloch, R. Zan, et al., Wide-area strain sensors based upon graphene-polymer composite coatings probed by Raman spectroscopy, *Advanced Functional Materials*, 24 (2014) 2865–2874.

37. S. Gong, W. Schwalb, Y. Wang, Y. Chen, Y. Tang, J. Si, et al., A wearable and highly sensitive pressure sensor with ultrathin gold nanowires, *Nature Communications*, 5 (2014) 1–8.

38. K. Dubbert, D. Völker, P. Apel, *Fact Sheet – Use of Nanomaterials in Coatings*, Federal Environment Agency, Dessau-Roßlau, Germany, 2014.

39. C. Li, X. Guo, G.S. Frankel, Smart coating with dual-pH sensitive, inhibitor-loaded nanofibers for corrosion protection, *Materials Degradation*, 5 (2021) 1–13.

4 Smart Coatings
Real-Time Applications in Bioimplant Materials

4.1 INTRODUCTION

The current century has been recognized as a "materials century". There are two consecutive reasons to believe this argument: (i) extensive use of naturally occurring materials and (ii) modification/doping/mixing of these materials in an alloy to serve the emerging needs of the society and eventually improve the quality of human life. In view of this, the field of bioimplant or bioinspired materials has shown generous progress in the past two decades due to their excellent functional properties such as biodegradability, antireflectivity, self-cleaning, and self-healing abilities. However, with ever increasing demands in miniaturization due to the growing interdependence of multidisciplinary fields, materials scientists started evolving around the development of auto-responsive coatings to protect the surface from biofouling and corrosion in biomedical/implant materials that are known as bioinspired smart coatings (BSCs).

The exploration of BSCs requires a profound knowledge of biological principles in conjunction with knowledge of medical, environmental, aerospace, and automobile fields (depends on the design of BSCs for a specific application). Besides this, one of the key aspects while designing the BSCs is the living behaviour or response of companion organisms from the underlying environment. Applying BSCs on bioimplant materials is expected to enhance biodegradability and bioactivity, thus fitting them with the human body environment [1–3]. These BSCs are also designed to expedite the targeted drug deliveries in treating various diseases like cancer and tumours (a detailed discussion is provided in Chapter 5). Furthermore, the biocompatibility of these BSCs with the human body is also an important consideration while applying these BSCs on bioimplant materials. For example, these BSCs have to work in a variable thermal environment, wsherein the temperature of the human/animal body changes from person to person.

Of the several BSCs, shape memory alloys/polymers (SMA/Ps), nanoceramic, and nanobioceramic BSCs are quite popular and promising candidates for the aerospace, biomedical, automotive, and robotics applications owing to their excellent pseudoelastic effect (in SMA/Ps), self-generating, self-healing, and self-cleaning characteristics. Below we shed light on these BSCs, particularly for biomedical applications. The various synthesis and applying techniques in the above-metioned BSCs are already well-documented and beyond the scope of the

DOI: 10.1201/9781003200635-4

current book. However, we mainly emphasize the underlying protection mechanisms offered by these coatings and their advancement in potential applications.

4.2 STIMULI RESPONSE OF SCS: APPLICATIONS IN BIOIMPLANT MATERIALS

4.2.1 SHAPE MEMORY ALLOYS/POLYMER SCs

This section discusses the underlying protection mechanisms offered by BSCs in bioimplant materials. In 1932, Arne Ölander [4] found some naturally occurring crystals of gold and cadmium that can recover their original shape when heated through a certain transformation temperature, T_s. This eventually led to the birth of the "shape memory effect (SME)", which exhibits pseudoelasticity. The first work on the application of SME in the medical field was reported in 1941, where Vernon and Vernon [5] introduced pseudoelasticity in polymers. Since then, a great deal of interest has been found to introduce this SME in Nickel–Titanium (Ni–Ti) [6], shape memory polymers (SMPs) [5,7], shape memory gels [7], and shape memory ceramics [8] for biomedical, robotics, aerospace, and automotive applications. Of all the SMAs, Ni–Ti is the potential candidate for surgical devices, bioimplant stents, and various bioimplant materials owing to their excellent shape regaining ability and biocompatibility [9,10], while SMPs represent a class of stimuli response SCs [11,12].

The preparation of SMA/Ps involves conventional and advanced techniques such as thermal spray coating, spin coating, dip coating, painting, magnetron sputtering, vacuum evaporation, physical and chemical vapour deposition, pulsed laser deposition, and ion-beam evaporation [1]. Recently, studies have reported the ability of this SME in stimuli-responsive SCs for the protection of the surface of bioimplant materials [13–15]. The protection strategy involves incorporating polymer healing agents (discussed in the previous chapters) and extrinsic corrosion inhibitors *via* microencapsulation. One of the key mechanisms behind the stimuli response of self-healing/repairing SMA/Ps SCs is triggering *via* pH, temperature, and light (triggering mechanisms are discussed in the previous chapters) [16–18].

The role of biocompatibility while designing SMA/Ps SCs and thin films for the biomedical implant is of paramount importance. For instance, owing to their superelastic properties, excellent biocompatibility, and good anticorrosion characteristics, Ni–Ti SMAs are introduced in biomedical applications. The progress made in the last decade to produce the thin film of Ni–Ti for biomedical devices is notable. These thin films can be applied/deposited on the external surfaces of medical or bioimplant materials/devices placed for probing blood composition (chemistry), biochemistry, replacement of bones and joints, blood flow measuring devices and are expected to provide superior protection, self-healing, and self-generating characteristics when the need arises. For instance, Biswas and co-workers [19] have successfully demonstrated the ability of SMPs SCs made of polyurethane nanohybrid. This polyurethane nanohybrid SCs were prepared *via in situ* polymerization of aliphatic diisocyanate, ester polyol, and a chain extender in the presence of two-dimensional platelets. *In vivo* studies on albino rats have indicated that the prepared SCs exhibit SME, which can heal the lips of the wound *via* self-tightening mechanisms. The rat temperature was the key triggering mechanism used during the application of these coatings. The biodegradability of these coatings was also confirmed in the temperature zone 35°C–40°C, thus extending its applications to self-expanding stents. In another work, the SME in Ni–Ti film has effectively demonstrated the permanent and temporary clipping of bold vessels for the medical devices used in arterial applications [20]. The one-way and two-way SME in Ni–Ti was successfully obtained *via* systematically annealing pristine Ni–Ti thin film (thermo-mechanical treatment) and subsequent phase transformation. Although not tested *in vivo*, the temperature of the human body is anticipated to be an external trigger for the functioning of the thin film. The other typical applications of these SMA/Ps SCs or thin films in biomedical devices are in the omni-pod system in insulin pump [21], sutures [22], eyeglass frames [23], orthodontic archwires [24], implant devices in penile surgery/treatment [25], and artificial joints [26].

A notable work of Huang and co-workers [27] demonstrated the applicability of thermo-responsive ester-based thermoplastic polyurethane (another type of SMP) for the possible positioning of medical devices inside the human body, indicating the biocompatibility of SMA/Ps SCs. The Joule heating effect was introduced in polyurethane SMP upon heating to obtain the pseudoelastic effect on the materials. Figure 4.1 demonstrates the active shape recovery in ~300 nm thick thermo-responsive polyurethane SMP after inducing SME. This designed SMP thin film offered excellent thermo-moisture responsive characteristics to cell and minimally invasive surgery applications.

The studies by Boccaccini et al. [28] reported the viability of polyetheretherketone (PEEK) and PEEK/bioglass coatings on Ni–Ti wires for protecting the Ni–Ti surface against corrosion in body fluid. These coatings were prepared by electrophoretic deposition and offered excellent adhesion to the underlying Ni–Ti substrate. The bioactive nature of bioglass is advantageous to improve the bonding between bones' soft tissues and joints, thereby excelling the healing abilities

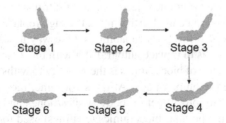

FIGURE 4.1 Schematic demonstrating the various stages of shape recovery (also SME) in the ~300 nm (not to the scale) thick thermo-responsive polyurethane SMP upon heating. (Originally adopted from W. Huang et al., *Journal of Materials Chemistry*, 20 (2010) 3367–3381 [27] and schematically redrawn.)

of SMA/Ps SCs. Although these SCs provided excellent multifunctional characteristics and served the purpose of stimuli response in biomedical devices at the laboratory level, their practical applications at a commercial scale are still a major challenge. This could be due to the issues related to their bioactivity, corrosion protection characteristics of layer and its effectiveness, and issues associated with the surface roughness of the protective layer. We shed light on the issues related to these aspects in Section 4.3.

4.2.2 NANOCERAMIC AND NANOBIOCERAMIC SCs

Ceramics are nonmetallic inorganic material processed at high temperatures. It possesses high hardness and strength, low ductility. Any ceramic could be comprised of both metal and nonmetal. However, the end product is nonmetallic in nature. An advent in ceramic engineering opened up new pathways to synthesize ceramics at micro- and nanoscale, thereby producing the ceramics/ceramet coatings or thin films. A nanoceramic is a material capable of producing these coatings in thin film form. Nanoceramic coatings are promising candidates, especially against high wear and temperature. Interestingly, the dielectric, piezoelectric, pyroelectric, ferromagnetic, and electromagnetic effects can be introduced in the coatings made of hard nanoceramics and are advantageous in coupled loading environments. The main difference between nanocomposite and nanoceramic is that nanoceramic is the sub-component of the nanocomposite. A nanocomposite comprises a nanoceramic- and nanopolymer-embedded metallic matrix, while nanoceramic is only a repeated geometry of metal and nonmetal atoms with a periodic arrangement [2]. An interesting feature of these nanoceramic coatings, which makes them suitable for bioimplant materials, is their high corrosion resistance even in the aggressive environment (ceramics being nonmetallic, generally do not corrode).

The number of bone replacement/repair/implant surgeries happening across the globe is increasingly apparent. Until the past decade, the medical community often practised using autografts, which can pick the bone from one part of the body and place it on another. This practice is followed by the bone surgeon (practitioner) for bone regeneration. However, one of the downsides of autografts is that they could create immune rejection and viral transmission in patients, and sometimes, it may lead to death. Furthermore, the high replacement (surgery) cost with limited supply also limits its capabilities. One way to alleviate this tedious and expensive process is to replace autografting with bioceramics [2].

An intriguing feature of bioceramics is their ability to withstand the aggressive environment inside the human body. At the same time, these nano/bioceramics also have a less deleterious effect on body tissues/bones and human health. On this front, some of the popular bioceramics (coatings) used for biomedical applications are hydroxyapatite (HA), tricalcium phosphate, dicalcium phosphate, and octcalcium phosphate [2,29]. Of all these, HA is emerging out to be a potential candidate for bone regeneration and replacement owing to its similar structure to human bone [2,29,30]. The application of nano/bioceramics either in bulk or nanocoating form could provide a breakthrough to the biomedical community as it reduces the bone replacement/surgical cost and eliminates the need for growth hormones and biomolecules.

The typical synthesis techniques to prepare HA-incorporated nano/bioceramics SCs are sol–gel processing, ultrasonic technique, and hydro-thermal processing. Interestingly, calcium phosphate (Ca-P) nano/bioceramics SCs can also be synthesized *via* sintering [2]. One of the key characteristics of bioceramic SCs from their applications in medical devices or bone replacements point of view is their bioactivity. The bioactivity of bioceramics in the medical sense can be defined as the systematic activation of pores on the surface of ceramics (or coatings) to accommodate the bone-growth proteins, osteoblasts, and other biomolecules. This bioactivity of bioceramics (or coatings) can be enhanced *via* controlling the pore size, shape, and density on the surface. For instance, the decrease in pore size and the increase in the number of pores per surface are expected to enhance the bioactivity of bioceramics [31].

In the current scenario, the biomedical community is also fascinated by the application of nano/bioceramics (either in bulk or coatings form) in hip implants. Due to their high fretting and fracture resistance and sturdiness, these nano/bioceramics can replace autografting in bone replacement/regeneration surgery. Some commonly used materials on this front are ceramics of titanium and zirconium alloys such as alumina–titania (AT) and yttria-stabilized zirconia (YSZ). The mixture of AT and YSZ is also preferred in many instances against high wear in titanium alloys [2,32]. Of all these, nano/bioceramics of titanium alloys are quite popular due to their elastic modulus, nearly equal to human bones. Besides this, they also possess good biocompatibility and high specific strength (i.e., high strength to weight ratio), which is of paramount importance from the biomedical applications point of view. The profound review of the literature indicated that the field of nano/bioceramics has extended its wing in daily/domestic human life. However, maintaining their bioactivity at smaller length scales due to difficulties in control of pores during synthesis/fabrication is an open challenge. Below, we brief out the key issues related to nano/bioceramics in biomedical materials/devices.

4.3 CURRENT ISSUES AND FUTURE PERSPECTIVES

The field of SCs in the past two decades has opened up new pathways to tailor the functional performance of bioimplant materials. Therefore, it is believed that the business in SCs for biomedical applications will show a projected market of $2–4 billion (US dollars) in the next 5 years [1–3]. Therefore, it can be construed that many unfolded applications of these coatings are still researched at the laboratory level and are yet to occupy the commercial market. For instance, owing to their applicability in expandable stents due to SME (or pseudoelasticity), SMAs/Ps are also being researched for their applications in controlled drug delivery. However, the need for the timely response of these SCs in implant surgeries necessitates scientists to develop optimal combinations of constituents and relative triggering mechanisms. On this front, one of the key triggering mechanisms is surrounding temperature (say body temperature in this case), which may vary from human to human and animal to animal. For example, the SME effect in SMA/Ps SCs is very sensitive to the change in temperature when viewed from the nanoscale (i.e., in thin films) and may disturb the shape recovery. This can eventually cause an adverse effect on human health. Therefore, the biocompatibility aspects should be of utmost importance while designing SMA/Ps SCs for bioimplant materials. Similarly, nano/bioceramics SCs possess excellent bioactivity. However, control of pores (i.e., shape and density) on the surface is of primary importance to accommodate the secreted bone proteins and biomolecules during bone and hip replacement. Furthermore, the thermo-responsive characteristics of nano/bioceramics SCs have not been fully exploited. Therefore, it can be concluded that the above issues pave the realming shift in the SCs market in the next decade.

The future prospects of the above SCs are anticipated to be based on the following aspects: (i) multifunctional characteristics (i.e., self-healing, thermo-resistive, and self-cleaning), (ii) ease of operation and applications, and (iii) improvement in bioactivity and biocompatibility. For instance, recent studies by Makhlouf and co-workers [33–38] have exploited the potential application of air-jet spinning techniques to prepare single step nano/bioceramic SCs for bioimplant materials. This eliminates the multi-step synthesis and fabrication and concurrently trims down the production cost. It would be prudent that these coatings can be produced on a commercial scale to serve the emerging needs of the biomedical community. In addition to this, the dual air-jet spinning technique is also employed by their group, which is novel, cost effective, and less tedious [39,40]. The enhancement in response time when the need arises is also one of the key issues in nano/bioceramic SCs. In view of this, catalyst materials can expedite the triggering and eventually the response time. The incorporation of nanoglass in nano/bioceramic SCs could enhance the bioactivity and subsequently the multifunctional characteristics. All in all, the current research trends in SCs for biomedical applications sit squarely in the nexus of numerous opportunities.

4.4 CONCLUSIONS

It is no wonder that SCs have occupied almost 70% market in biomedical applications when we think of applying a protective layer on biomedical materials/ devices. Makhlouf and co-workers [1–3] have reported that there are ~50,000–70,000 implant surgeries happening across the globe every day, and the count may increase in the next 5 years or so. The key challenge in front of the biomedical community is a development/application of a layer on biomedical materials/ devices that can offer stimuli response mechanisms such as self-healing, self-cleaning, and self-generating *via* intrinsic triggering (say the change in pH and body temperature) along with enhanced bioactivity and biocompatibility.

In the present chapter, we provided a systematic and concise discussion on different types of SCs used to protect the surface in bioimplant materials and their (multi)functional characteristics along with triggering and protection mechanisms. First and foremost, SMA/Ps SCs have effectively offered protection in various medical applications such as sutures, eyeglass frames, orthodontic archwires, implant devices in penile surgery/treatment, and artificial joints. The one-way and two-way SME in SMA/Ps SCs can be effectively used to restore the original shape of the coating material and eventually clipping bold vessels for the medical devices used in arterial applications. On the other hand, nano/bioceramic SCs are viable in bone replacement and hip implant surgeries, thereby eliminating the conventional tedious autografting process. The key triggering mechanisms which are common in all the SCs used for bioimplant applications are changes in pH and surrounding temperature (say body temperature in this case). Although these SCs are widely researched at a laboratory scale, their practical applications at a commercial scale are of utmost importance in the upcoming time. A large number of trials are underway to enhance the biocompatibility and bioactivity of these SCs to suit them for implant surgeries. Nevertheless, these SCs have made impressive progress in the automotive, aerospace, and medical sectors.

REFERENCES

1. A.S.H. Makhlouf, N.Y. Abu-Thabit, D. Ferretiz, Shape-memory coatings, polymers, and alloys with self-healing functionality for medical and industrial applications, In: Makhlouf, A.S.H., Abu-Thabit, N.Y. (eds.) *Advances in Smart Coatings and Thin Films for Future Industrial and Biomedical Engineering Applications*, 1st Ed., pp. 335–358. Elsevier (2020).
2. A.S.H. Makhlouf, E. Guerrero, Toward smart and durable nanoceramic and nanobioceramic coatings and their medical and industrial applications, In: Makhlouf, A.S.H., Abu-Thabit, N.Y. (eds.) *Advances in Smart Coatings and Thin Films for Future Industrial and Biomedical Engineering Applications*, 1st Ed., pp. 383–403. Elsevier (2020).
3. A.S.H. Makhlouf, R. Rodriguez, Bioinspired smart coatings and engineering materials for industrial and biomedical applications, In: *Advances in Smart Coatings and Thin Films for Future Industrial and Biomedical Engineering Applications*, 1st Ed., pp. 335–358. Elsevier (2020).
4. A. Ölander, An electrochemical investigation of solid cadmium gold alloys, *Journal of American Chemical Society*, 54(10) (1932) 3819–3833.
5. L.B. Vernon, H.M. Vernon, Process of manufacturing articles of thermoplastic synthetic resins, Google Patents, 1941.
6. W.J. Buehler, J. Gilfrich, R. Wiley, Effect of low-temperature phase changes on the mechanical properties of alloys near composition TiNi, *Journal of Applied Physics*, 34(5) (1963) 1475–1477.
7. F. Pilate, A. Toncheva, P. Dubois, J.M. Raquez, Shape-memory polymers for multiple applications in the materials world, *European Polymer Journal*, 80 (2016) 268–294.
8. K. Uchino, Antiferroelectric shape memory ceramics, Actuators, Multidisciplinary Digital Publishing Institute, 2016.
9. F. Auricchio, E. Boatti, M. Conti, Chapter 11: SMA biomedical applications, In: L. Lecce, A. Concilio (eds.), *Shape Memory Alloy Engineering*, pp. 307–341. Butterworth-Heinemann, Boston (2015).

10. C. Craciunescu, A.S. Hamdy, The effect of copper alloying element on the corrosion characteristics of TiNi and ternary Ni-Ti-Cu meltspun shape memory alloy ribbons in 0.9% NaCl solution, *International Journal of Electrochemical Science*, 8(8) (2013) 10320–10334.
11. Y. Guo, Z. Lv, Y. Huo, L. Sun, S. Chen, Z. Liu, C. He, X. Bi, X. Fan, Z. Yu, A biodegradable functional water-responsive shape memory polymer for biomedical applications, *Journal of Materials Chemistry B*, 7(1) (2019) 123–132.
12. Q. Zhao, H.J. Qi, T. Xie, Recent progress in shape memory polymer: new behavior, enabling materials, and mechanistic understanding, *Progress in Polymer Science*, 49–50 (2015) 79–120.
13. J.K. Lee, X. Liu, S.H. Yoon, M.R. Kessler, Thermal analysis of ring-opening metathesis polymerized healing agents, *Journal of Polymer Science B: Polymer Physics*, 45(14) (2007) 1771–1780.
14. X. Liu, J.K. Lee, S.H. Yoon, M.R. Kessler, Characterization of diene monomers as healing agents for autonomic damage repair, *Journal of Applied Polymer Science*, 101(3) (2006) 1266–1272.
15. X.K. Hillewaere, F.E. Du Prez, Fifteen chemistries for autonomous external self-healing polymers and composites, *Progress in Polymer Science*, 49 (2015) 121–153.
16. S.H. Cho, S.R. White, P.V. Braun, Self-healing polymer coatings, *Advanced Materials*, 21(6) (2009) 645–649.
17. E. Shchukina, D.G. Shchukin, Nanocontainer-based active systems: from self-healing coatings to thermal energy storage, *Langmuir*, 35 (2019) 8603–8611.
18. N.Y. Abu-Thabit, A.S. Hamdy, Stimuli-responsive polyelectrolyte multilayers for fabrication of self-healing coatings – a review, *Surface and Coatings Technology*, 303 (2016) 406–424.
19. A. Biswas, A.P. Singh, D. Rana, V.K. Aswal, P. Maiti, Biodegradable toughened nanohybrid shape memory polymer for smart biomedical applications, *Nanoscale*, 10(21) (2018) 9917–9934.
20. E. Ryklina, A. Korotitskiy, I. Khemelevskaya, S. Prokoshkin, K. Polyakova, A. Kolobova, M. Soutorine, A. Chernov, Control of phase transformations and microstructure for optimum realization of one-way and two-way shape memory effects in removable surgical clips, *Materials Design*, 136 (2017) 174–184.
21. H.C. Zisser, The Omni Pod insulin management system: the latest innovation in insulin pump therapy, *Diabetes Therapy*, 1(1) (2010) 10–24.
22. R. Seguchi, N. Ishikawa, T. Tarui, T. Horikawa, T. Ushijima, G. Watanabe, A novel shape-memory monofilament suture for minimally invasive thoracoscopic cardiac surgery, *Innovations* 14 (2019) 55–59.
23. M. Salerno, K. Zhang, A. Menciassi, J.S. Dai, A novel 4-DOF origami grasper with an SMA-actuation system for minimally invasive surgery, *IEEE Transactions Robotics*, 32(3) (2016) 484–498.
24. Y.F. Liu, J.L. Wu, S.L. Song, L.X. Xu, J. Chen, W. Peng, Thermo-mechanical properties of glass fiber reinforced shape memory polyurethane for orthodontic application, *Journal of Materials Science: Materials Medicine*, 29(9) (2018) 148.
25. B.V. Le, K.T. Mcvary, K. McKenna, A. Colombo, Use of magnetic induction to activate a "Touchless" shape memory alloy implantable penile prosthesis, *Journal of Sexual Medicine*, 16(4) (2019) 596–601.
26. K. Takashima, J. Rossiter, T. Mukai, McKibben artificial muscle using shape memory polymer, *Sensors and Actuators A: Physics*, 164(1–2) (2010) 116–124.
27. W. Huang, B. Yang, Y. Zhao, Z. Ding, Thermo-moisture responsive polyurethane shape memory polymer and composites: a review, *Journal of Materials Chemistry*, 20 (2010) 3367–3381.

28. A.R. Boccaccini, C. Peters, J.A. Roether, D. Eifler, S.K. Mishra, E.J. Minay, Electrophoretic deposition of polyetheretherketone (PEEK) and PEEK/Bioglass coatings on NiTi shape memory alloy wires, *Journal of Materials Science*, 41(24) (2006) 8152–8159.

29. Y. Hong, H. Fan, B. Li, B. Guo, M. Liu, X. Zhang, Fabrication, biological effects, and medical applications of calcium phosphate nanoceramics, *Materials Science and Engineering R: Reports*, 70(3–6) (2010) 225–242.

30. A. Abdal-hay, A.S. Hamdy, J. Lim, Facile preparation of titanium dioxide micro/nanofibers and tubular structures by air jet spinning, *Ceramics International*, 40(10A) (2014) 15403–15409.

31. A. Abdal-hay, A.S. Hamdy Makhlouf, P. Vanegas, Chapter 4: A novel approach for facile synthesis of biocompatible PVA-coated PLA nanofibers as composite membrane scaffolds for enhanced osteoblasts proliferation, In: A.S.H. Makhlouf, D. Scharnweber (eds.), *Handbook of Nanoceramic and Nanocomposite Coatings and Materials*, pp. 87–112. Elsevier (2015).

32. M. Semlitsch, Titanium alloys for hip joint replacements, *Clinical Materials*, 2(1) (1987) 1–13.

33. A. Abdal-hay, K.H. Hussein, L. Casettari, K.A. Khalil, A.S. Hamdy, Fabrication of novel high performance ductile poly(lactic acid) nanofiber scaffold coated with poly (vinyl alcohol) for tissue engineering applications, *Materials Science and Engineering C*, 60 (2016) 143–150.

34. A. Abdal-hay, A.S. Hamdy, M.Y. Abdellah, J. Lim, In vitro bioactivity of implantable Ti materials coated with PVAc membrane layer, *Materials Letters*, 126(7) (2014) 267–270.

35. A. Abdal-hay, P. Vanegas, A.S. Hamdy, F.B. Engel, J. Lim, Preparation and characterization of vertically arrayed hydroxyapatite nanoplates on electrospun nanofibers for bone tissue engineering, *Chemical Engineering Journal*, 254(10) (2014) 612–622.

36. A. Abdal-hay, A.S. Hamdy, K.A. Khalil, J. Lim, A novel simple in situ biomimetic and simultaneous deposition of bonelike apatite within biopolymer matrix as bone graft substitutes, *Materials Letters*, 137(12) (2014) 260–264.

37. A. Abdal-hay, A.S. Hamdy, Y. Morsi, K.A. Khalil, J. Lim, Novel bone regeneration matrix for next-generation biomaterial using a vertical array of carbonated hydroxyapatite nanoplates onto electrospun Nylon 6 nanofibers, *Materials Letters*, 137(12) (2014) 378–381.

38. A. Abdal-hay, A.S. Hamdy, K.A. Khalil, Fabrication of durable high performance hybrid nanofiber scaffolds for bone tissue regeneration using a novel, simple in situ deposition approach of polyvinyl alcohol on electrospun Nylon 6 nanofibers, *Materials Letters*, 147(5) (2015) 25–28.

39. A. Abdal-hay, A.S. Hamdy, F.B. Engel, J. Lim, A novel simple one-step air jet spinning approach for deposition of poly(vinyl acetate)/hydroxyapatite composite nanofibers on Ti implants, *Materials Science Engineering C*, 49(4) (2015) 681–690.

40. A. Abdal-hay, A. Memic, K.H. Hussein, Y.S. Oh, M. Fouad, F.F. Al-Jassir, Rapid fabrication of highly porous and biocompatible composite textile tubular scaffold for vascular tissue engineering, *European Polymer Journal*, 96 (2017) 27–43.

5 Challenges in Clinical Translations

5.1 INTRODUCTION

The word "Clinical Translations" involves human beings. Readers might be wondering how does it include people like us? We will split the word "Clinical Translations" into two parts for a better understanding. The term "Clinical" refers to the persons undergoing clinical investigations who voluntarily participate in clinical trials to understand the science behind any disease, its diagnosis, and modifications/developments in treatments. On the other hand, the term "Translations" in the medical sense refers to direct applications of medical discoveries (invented in medical research labs) on human health and these best practices are further adopted for the welfare of the entire community. With the advent of this field, advanced/nanoparticulate nanomaterials (NNMs) have received significant attention because they can directly impact the (bio)medical field and eventually on human health *via* the diagnosis, treatment, and prevention of diseases.

While these NNMs are the obvious choice of materials scientists to effectively reduce the toxicity and increase the therapeutic index [1], therefore their surface functionalization and protection become increasingly apparent as these materials continuously have to work in an aggressive environment. In view of this, smart coatings (SCs) and related triggering mechanisms are of paramount importance. For instance, nanoencapsulation (discussed in the previous chapters) protects the fragile layer/compound, which degrades in the biological environment. This nanoencapsulation is also expected to provide better solubilization [1–5]. Further, the triggered release of nanomaterials in SCs has been proven beneficial for improving the therapeutic index. Drug molecules can be efficiently delivered to the targeted sites without affecting the healthy sites in the human body [6–8].

Keeping the importance of the above aspects in view, we discuss the various protection mechanisms offered by SCs in clinical translations. The discussion presented below is only limited to the related triggering and targeted mechanisms, thus synthesis and applying techniques of SCs for the above application are beyond the scope of present work, as these aspects are already very well documented in the previous literature.

5.2 SMART COATINGS (SCs): PROTECTION MECHANISMS IN CLINICAL TRANSLATIONS

The SCs are often used for improving the targeted drug delivery in clinical translations. However, at the same time, they should essentially not affect the healthy

DOI: 10.1201/9781003200635-5

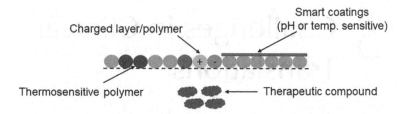

FIGURE 5.1 Schematic representation of triggered release of NNM from SCs to therapeutic sites. (Adopted and modified with permission under CC BY 4.0 License from S. Hua et al., *Frontiers in Pharmacology*, 9 (2018) 790(1)–790(14) [1].)

sites/tissues. Owing to their multifunctional characteristics and ability to encapsulate the NNM, these SCs find a unique niche in clinical development and clinical use applications for treating cancers and tumours [9]. However, from the past decade, the extensive use of SCs for targeting non-cancer diseases such as inflammatory diseases, which typically includes asthma, inflammatory bowel disease, multiple sclerosis, diabetes, and neurodegenerative diseases, is increasingly apparent [10]. Enhanced permeability and retention (EPR) and passive accumulation of NNM in SCs appear to be important reasons behind the extensive use of NNM-incorporated SCs in clinical use [1]. The EPR effect utilizes the active release of NNM in tissues, while the pathological properties of tissues allow triggered NNM to accumulate at pathological sites, known as passive accumulation. Figure 5.1 demonstrates the triggering of NNM from SCs at the targeted site, and the protection is based on EPR and passive accumulation mechanism. The application of SCs in clinical translations is mainly based on two mechanisms and is discussed briefly below.

5.2.1 ACTIVE TARGETING

Active triggering utilizes ligands such as antibodies and sugar moieties to ease the localization of target cells by enhancing their uptake. These ligands are expected to agglomerate on the surface of NNM-incorporated SCs by physicochemical means [11]. The schematic in Figure 5.2 demonstrates the active triggering of ligands to therapeutic sites. Interestingly, these ligands on the surface of NNM-incorporated SCs have the potential of site-specific delivery to designated cells *in vivo*, which eventually enhances the adhesion to the site-specific molecules [1,7,12]. For instance, cancer cells, tumour endothelium, and stroma cells are target-specific sites in targeted cancer prevention. Several studies [13–16] have been in the past argued that NNM-incorporated SCs are capable of the active release with significant enhancement of accumulation of NNM at the targeted sites compared to non-targeted sites. This eventually indicates that the ERP is dominant over the passive triggering/accumulation of NNM in SCs.

While most of the studies have reported the EPR and passive accumulation as dominating mechanisms in triggering/active releasing of NNM, the observations

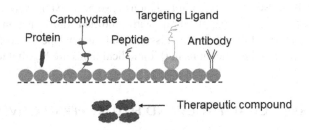

FIGURE 5.2 Schematic representation of targeted ligand drug delivery *via* triggered release of NNM from SCs to therapeutic sites. (Adopted and modified with permission under CC BY 4.0 License from S. Hua et al., *Frontiers in Pharmacology*, 9 (2018) 790(1)– 790(14) [1].)

of Kirpotin and co-workers [17] revealed no significant difference between accumulation of NNM at targeted sites and non-targeted sites. Similar levels of tumour tissues were observed in the breast cancer xenografts models. Currently, investigations on the applications of SCs as an active triggering medium are a hot and timely topic. More research and a strong network of medical researchers with coatings scientists is indeed a requirement. This also requires to excel in the interdisciplinary approach of (bio)medical, bioscience, materials, and coatings technology fields. Besides this, the active triggering mechanism is also of vital importance considering triggered drug release and is explained briefly below.

5.2.2 Triggered Release

Owing to their multifunctionality (i.e., endogenous or exogenous triggering and delivery), the triggered release of NNMs from SCs to target-specific sites is gaining significant attention of researchers. The NNM from SCs to a specific site (say tumour) can be triggered *via* endogenous or exogenous stimuli. While the endogenous response of SCs in triggered drug release is based on the change in pH values, the influence of enzymes, and redox activities in the site-specific microenvironment, the latter is indicative of triggering *via* change in temperature excitation magnetic field and ultrasound [1]. Of all the external triggering agents, the use of thermosensitive liposomal doxorubicin (Figure 5.1) is perhaps the most promising one owing to its superior temperature-sensitive characteristics over Doxil® [18]. Several studies [19–21] have been in the past reported the applicability of thermosensitive liposomal doxorubicin in an *in vivo* mode in non-resectable hepatocellular carcinoma. Incorporating thermosensitive liposomes with temperature-sensitive lipids (for example, polymers such as poly(N-isopropylacrylamide)) is expected to enhance the stability of NNM in SCs as well as help them to maintain their contents at varying temperatures. Interestingly, this feature also allows them to change their phase upon heating, making them favourable for triggering or releasing NNM from SCs [18–21]. The triggered release of NNM with the incorporation of additional agents can also be used for probing the biodistribution and target accumulation [1,18–21].

Although these SCs for clinical translations have been extensively researched at a laboratory scale, their applications with modified multifunctionalities are timely topic. In view of this, the current issues and future perspectives of SCs considering their potential applications in (bio)medical fields are highlighted below.

5.3 CURRENT CHALLENGES AND FUTURE PERSPECTIVES

In one of the review articles, Hua and co-workers [1] have reported that the real-time application of stimuli-responsive NNM or SCs in clinical translations is a time consuming and costly affair. The application of SCs in clinical translations is more tedious and complex than conventional coating technologies. For instance, key issues such as biocompatibility and biological hurdles, cost-effective implementation, and passing government regulations are the topmost priorities of the coating community. This indeed requires coordination between interdisciplinary fields, including materials and coatings science, bioscience, therapeutic science, and biomedical engineering.

From the biological and biocompatibility viewpoint, understanding the relationship between bioscience, bioengineering, and biotechnology is of primary importance. For the successful applications of NNM-incorporated SCs, the knowledge of delivery system characteristics and *in vivo* behaviour in humans/animal is necessary. In view of this, Hare and co-workers [9] have proposed disease-driven approach to achieve the maximum efficacy of NNM-incorporated SCs, and subsequently, the pathophysiological changes in disease. Furthermore, the dissociative nature of the human disease is also an important barrier that needs to be considered for improving the triggered release of NNM from SCs to a target-specific site. On the other hand, this approach should also consider the minimum accumulation of NNM at non-targeted (healthy) sites. Although the research at the laboratory level for applying NNM-incorporated SCs is proven to overcome the biological barriers, the comprehensive understanding of inter-relations between patient health conditions, biological heterogeneity, and drug delivery in clinical translations is still far from complete. Furthermore, most of the research on the above aspects is focused on cancer targeting/prevention applications, and many clinical translations remain unexploited. In view of this, ligand-targeted site delivery can be advantageous for enhanced accumulation of NNM at the targeted site.

Another important challenge in applying these coatings in clinical translations is the overall safety of human/animal life (biocompatibility). Detailed knowledge of toxicology is indeed a requirement. To safeguard the use of these coatings, the knowledge of drug product, triggering agent, and coating formulation is required. This typically includes (i) changes happening during the drug delivery at targeted sites, (ii) changes reflecting from the impurities associated with NNM or polymer agent, (iii) changes arising from the formulation and synthesis of NNM, and (iv) changes occurring in base or parent material. *In/ex vivo* and *in/ex vitro* protocols should be followed for regulating such practices [22].

The development of any material or technology across the globe is governed by the government rules and regulations of the respective country. Considering the applications of NNM-incorporated SCs, they can excel in the (bio)medical and pharmaceutical market. However, commercialization of these coatings in clinical translations should not drive away from the purpose of safeguarding human health and the quality of coatings. This commercialization of NNM-incorporated SCs is governed by the government protocols for various manufacturing/synthesis techniques, quality assurance, safety, and intellectual property rights [23,24]. This eventually creates a large gap between scientific research and standard manufacturing techniques. For example, most of the SCs use polymer agents. However, the efficacy of SCs is based on polydispersity and conjugate chemistry [1,23,24]. Human/animal health is expected to respond to these polymers (or polymer agents) differently, which further influences the quantity of dose and dose frequency. In simple words, there is an urge to constitute a regulatory body/organization, which can provide the regulatory guidelines/framework for using polymer or polymer agents in SCs for clinical translations.

Complexities associated with drug release and drug delivery are also an important challenge for coatings and the biomedical community, which eventually limit the uses of NNM-incorporated SCs for clinical use. This typically includes different doses and response times for drug delivery. These two aspects could be ultimately connected to the compatible formulation of SCs for clinical translations. Therefore, in view of this, simplification in formulation and applying techniques of SCs is an important consideration, particularly for clinical use. This will eventually trim down the manufacturing cost with improved efficacy. Future studies need to be focused on applications of SCs in ligand-targeted deliveries and modifications in translation findings of animals to suit/excel them for human health.

In summary, the applications of NNM-incorporated SCs for clinical translation are at the primary stage. Large-scale laboratory research is currently going on on animals such as rats, mice, and small/mini creatures, which eventually limits the applications of SCs due to size and dose constraints. It would be prudent to translate this research in future to human health.

5.4 CONCLUSIONS

In the present chapter, we have briefly discussed the potential applications of NNM-incorporated SCs in clinical translations. More emphasis is given on the protection and triggering mechanisms offered by NNM-incorporated SCs. Since the real-time applications of these coatings in the (bio)medical field and clinical use are at the development stage (extensive research is still going on to make them suitable for real-time applications), the current challenges and issues related to these coatings are also briefly outlined.

The underlying mechanisms such as active release and active triggering of NNM are being extensively researched in the current scenario. Biocompatibility and biological hurdles, design cost, the stimuli response time of SCs, and government regulations are the key issues while designing the SCs for drug delivery applications. Therefore, the future of NNM-incorporated SCs lies in forming a regulatory framework for the cost-effective and efficient design for targeted drug delivery in clinical applications.

REFERENCES

1. S. Hua, M.B.C. de Matos, J.M. Metselaar, G. Storm, Current trends and challenges in the clinical translation of nanoparticulate nanomedicines: pathways for translational development and commercialization, *Frontiers in Pharmacology*, 9 (2018) 790(1)–790(14).
2. H.J. Kim, A. Kim, K. Miyata, K. Kataoka, Recent progress in development of siRNA delivery vehicles for cancer therapy, *Advanced Drug Delivery Review*, 104 (2016) 61–77.
3. M. Larsson, W.T. Huang, D.M. Liu, D. Losic, Local coadministration of gene-silencing RNA and drugs in cancer therapy: state of the art and therapeutic potential, *Cancer Treatment Review*, 55 (2017) 128–135.
4. D.K. Mishra, N. Balekar, P.K. Mishra, Nanoengineered strategies for siRNA delivery: from target assessment to cancer therapeutic efficacy, *Drug Delivery and Translational Research*, 7 (2017) 346–358.
5. M. Talekar, T.H. Tran, M. Amiji, Translational nano-medicines: targeted therapeutic delivery for cancer and inflammatory diseases, *AAPS J*, 17 (2015) 813–827.
6. E. Mastrobattista, G.A. Koning, G. Storm, Immunoliposomes for the targeted delivery of antitumor drugs, *Advanced Drug Delivery Review*, 40 (1999) 103–127.
7. S. Hua, Targeting sites of inflammation: intercellular adhesion molecule-1 as a target for novel inflammatory therapies, *Frontiers in Pharmacology*, 4 (2013) 127.
8. S. Hua, E. Marks, J.J. Schneider, S. Keely, Advances in oral nano delivery systems for colon targeted drug delivery in inflammatory bowel disease: selective targeting to diseased versus healthy tissue, *Nanomedicine*, 11 (2015) 1117–1132.
9. J.I. Hare, T. Lammers, M.B. Ashford, S. Puri, G. Storm, S.T. Barry, Challenges and strategies in anti-cancer nanomedicine development: an industry perspective, *Advanced Drug Delivery Reviews*, 108 (2017) 25–38.
10. L.S. Milane, M. Amiji, *Nanomedicine for Inflammatory Diseases*, New York, NY: CRC Press, (2017).
11. F. Danhier, To exploit the tumor microenvironment: since the EPR effect fails in the clinic, what is the future of nanomedicine? *Journal of Controlled Release*, 244(Pt A) (2016) 108–121.
12. M. Willis, E. Forssen, Ligand-targeted liposomes, *Advanced Drug Delivery Reviews*, 29 (1998) 249–271.
13. M. Ferrari, Nanovector therapeutics, *Current Opinions in Chemical Biology*, 9 (2005) 343–346.
14. A. Puri, K. Loomis, B. Smith, J.H. Lee, A. Yavlovich, E. Heldman, R. Blumenthal, Lipid-based nanoparticles as pharmaceutical drug carriers: from concepts to clinic, *Critical Reviews in Therapeutic Drug Carrier System*, 26 (2009) 523–580.
15. K. Riehemann, S.W. Schneider, T.A. Luger, B. Godin, M. Ferrari, H. Fuchs, Nanomedicine–challenge and perspectives, *Angewandte Chemie International Edition English*, 48 (2009) 872–897.
16. R. van der Meel, L.J. Vehmeijer, R.J. Kok, G. Storm, E.V. van Gaal, Ligand-targeted particulate nanomedicines undergoing clinical evaluation: current status, *Advanced Drug Delivery Reviews*, 65 (2013) 1284–1298.

17. D. Kirpotin, J.W. Park, K. Hong, S. Zalipsky, W.L. Li, P. Carter, C.C. Benz, D. Papahadijopoulos, Sterically stabilized anti-HER2 immunoliposomes: design and targeting to human breast cancer cells in vitro, *Biochemistry*, 36 (1997) 66–75.
18. D. Needham, G. Anyarambhatla, G. Kong, M.W. Dewhirst, A new temperature-sensitive liposome for use with mild hyperthermia: characterization and testing in a human tumor xenograft model, *Cancer Research*, 60 (2000) 1197–1201.
19. V.P. Torchilin, Multifunctional nanocarriers, *Advanced Drug Delivery Reviews*, 58 (2006) 1532–1555.
20. J. Shi, P.W. Kantoff, R. Wooster, O.C. Farokhzad, Cancer nanomedicine: progress, challenges and opportunities, *Nature Reviews in Cancer*, 17 (2017) 20–37.
21. K. Kono, Thermosensitive polymer-modified liposomes, *Advanced Drug Delivery Reviews*, 53 (2001) 307–319.
22. L. Accomasso, C. Cristallini, C. Giachino, Risk assessment and risk minimization in nanomedicine: a need for predictive, alternative, and 3rs strategies, *Frontiers in Pharmacology*, 9 (2018) 228.
23. S. Tinkle, S.E. McNeil, S. Muhlebach, R. Bawa, G. Borchard, Y.C. Barenholz, L. Tamarkin, N. Desai, Nanomedicines: addressing the scientific and regulatory gap, *Annals of the New York Academy of Sciences*, 1313 (2014) 35–56.
24. V. Sainz, J. Conniot, A.I. Matos, C. Peres, E. Zupancic, L. Moura, L.C. Silva, H.F. Florindo, R.S. Gaspar, Regulatory aspects on nanomedicines, *Biochemical and Biophysics Research Communications*, 468 (2015) 504–510.

6 Surface Functionalization by Smart Switchable Coatings
Applications to Marine and Medical Sectors

6.1 INTRODUCTION

Bacterial infection and subsequent series of undesirable effects on human life is a long-standing topic of great concern to the entire medical community. Sometimes colonization of such bacteria on bioimplant materials is serious and may lead to an increase in morbidity (or illness) and sometimes death [1]. On the similar lines, bacterial attachments on various equipment used in transportation, food packaging, and marine industries (or structures) have a series of undesired events such as biofouling, and subsequently, contamination of the surrounding environment, the short service life of the structure, and eventually increased maintenance and repair costs [1–3]. One way to alleviate the above issues is to apply active antibacterial coatings on the surface, which is more prone to bacterial attack. These antibacterial coatings are expected to work with two mechanisms: (i) active strategy and (ii) passive strategy. The former indicates a "kill bacteria" event using biocidal agents, while the latter prevents the contamination of surface from bacteria attack using antifouling materials [1,4,5]. However, one of the downsides of these coatings is their "single-action nature", which eventually poses an important issue related to their multifunctionality. For instance, notable studies of Wei and co-workers [1] and Yu et al. [5] have clearly outlined the limitations of single-acting biocidal antimicrobial coatings. The issue arises when a layer of killed bacterial accumulates on the surface, thereby producing inflammations and consequently affecting the killing efficiency. Therefore, developing such coatings which can switch the surface as and when needed with effective bacterial protection characteristics is still an important open area. The switchable surface essentially acts either way (i.e., active and passive), wherein active characteristics could help in bacteria-killing action. In contrast, passive characteristics must avoid the colonization of active or killed bacteria. Note that, to provide efficient switchable

DOI: 10.1201/9781003200635-6

characteristics, the two surface functions (i.e., bacteria killing and antifouling) should be compatible. Developing an antibacterial surface essentially involves one function to initiate the process and switch to another when the need arises [1,3–7].

Another important application of SSCs, which is of particular interest to the coatings community, is in the marine sector, particularly for maintaining the superwettability on the surface for the separation of oil and water during oily wastewater treatment. The increasing extraction of oil and gas from the ocean across the globe has created several hazardous impacts on the environment. For example, the increase in the proportion of oily wastewater in the sea during the production of oil (or petroleum products) has endangered the marine ecosystem and, eventually, human health. Thus far, the conventional oil and water separation treatments such as applying a high-flux composite coating [8], stainless steel mesh coated with polytetrafluoroethylene micro/nanostructures [9], hydrogel-coated stainless steel mesh [10], and using highly wettable porous materials [11–14] have been effectively implemented. However, some deficiencies such as high processing time, maintaining the wettability of the surface, and high equipment cost limit their uses in practical applications. To overcome the above limitations, the stimuli response characteristics of SCs can also be used for switching the surface (or changing the wettability) for oil and water separation during oily wastewater treatment in the oil and gas industries. The various SCs have been used to date for the applications mentioned above. In the below sections, we outline and summarize their applying techniques and functional characteristics, thus serving the purpose of the present chapter.

6.2 SMART SWITCHABLE COATINGS (SSCs): SYNTHESIS, APPLYING TECHNIQUES, AND TRIGGERING MECHANISMS

This section introduces some synthesis and applying/fabrication techniques of SSCs, which effectively switch the surface when the need arises. A detailed discussion is provided on the applying techniques and physical and chemical characteristics of SSCs to serve the purpose of the present chapter.

6.2.1 OVERVIEW OF APPLICATIONS TO MARINE SECTORS

One of the essential characteristics of the switchable surface is superhydrophobicity. The switchable surface should work according to the need of specific applications. For example, Bian et al. [15] have proposed a low-cost switchable system for separating oil and water by applying filter paper or a layer of zeolite coatings over a separation surface. As both the materials (filter paper and zeolite) are easily and readily available in nature (or day to day life), this eliminates the use of costly equipment/accessories for creating a switchable surface. This system works on a simple mechanism; as the oil/water mixture is poured on the surface of filter paper or zeolite, only water droplets are allowed to pass, thus leaving behind the oil droplets on a switchable surface. The superoleophobic characteristics

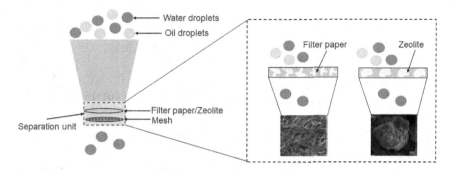

FIGURE 6.1 Schematic representing the process of oil/water separation from filter paper or zeolite. The surface morphologies of filter paper and zeolite are shown in insets. (Adopted and modified with permission under CC BY 4.0 License from H. Bian et al., *Frontiers in Chemistry*, 8 (2020) 1–7 [15].)

(i.e., oil-repellent tendency) of filter paper or zeolite may help to resist the oil droplets passing through the surface. The schematic in Figure 6.1 represents the above mechanism proposed by Bian et al. [15] along with the surface morphologies of filter paper and zeolite layer.

The chemical and physical characteristics such as porous nature (or porous membrane) of filter paper and tetrahedral skeleton of zeolite are helpful in the separation process. Filter paper is a permeable membrane of cotton fibre, while zeolite has SiO_4 and AlO_4 tetrahedral skeleton with wide gaps, which can further stack up in the material (as shown in Figure 6.1). The above oil/water separation method was found to be very effective as the measured separation efficiency of filter paper is ~96.07%, while that of the zeolite layer is ~97.23%.

In another attempt, Liu and co-workers [16] produced ammonia-induced SSCs with superwettable switchable surfaces to separate oil and water during oily wastewater treatment. The process of preparing ammonia-induced carnauba wax impregnated fluorosilane-modified titanium carboxylate hybrid gel SSCs (C-P-U@Ti) is illustrated in schematics in Figure 6.2. The tetrabutyl orthotitanate $Ti(OBu)_4$ (3 mL) was added in 5 mL ethanol with deionized water (1.6 mL) and undecanoic acid ($C_{10}H_{21}COOH$, UA). The prepared mixture was stirred for ~1.5 h at room temperature, followed by impregnation of the mixture (0.8 g carnauba wax + 20 mL toluene). The 1H,1H,2H,2H-Perfluorodecyltrimethoxysilane ($C_{13}H_{13}F_{17}Si$, PFDTMS) (1.95 mL) is then added, and the whole mixture is continuously stirred for 3 h. The typical substrate material used for applying the above coating is a porous sponge, which was then cleaned with detergent and subsequently washed with DI water and ethanol. The as-prepared coating was applied on the sponge *via* a simple dip technique and then dried in an oven at 65°C for 15 min, followed by exposure to an ammonium vapour environment.

The representative scanning electron microscope (SEM) images (Figure 6.3) revealed the porous nature of pristine sponge with uniform and smooth morphology, while NH_3-C-P-U@Ti coating offers a greater coverage to all the pores with

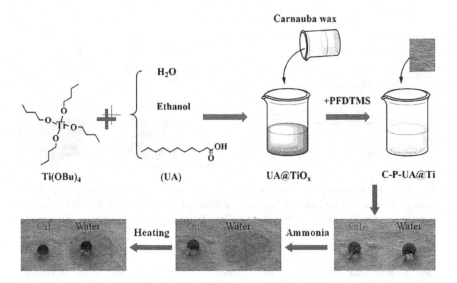

FIGURE 6.2 Schematic illustration of the fabrication process of superamphiphobic C-P-U@Ti SSC and its reversible switchable superwettable surface. (Adopted and reprinted with permission from W. Liu et al., *Journal of Environmental Chemical Engineering*, 8 (2020) 104164 [16].)

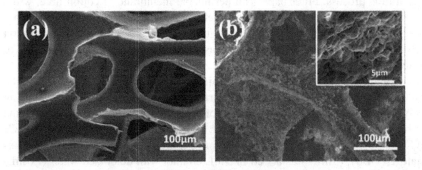

FIGURE 6.3 The morphologies of (a) pristine sponge and (b) sponge coated with C-P-U@Ti SSC. (Adopted and reprinted with permission from W. Liu et al., *Journal of Environmental Chemical Engineering*, 8 (2020) 104164 [16].)

a comparatively rough surface. The increased surface roughness of coated sponge is expected to enhance the hydrophobic/philic characteristics. The produced coatings also exhibited superior hydrophobicity and superoleophobicity when immersed in an oil (diesel)/water mixture. The pristine sponge is expected to submerge in the mixture, while the coated sponge is observed to be floating in the same mixture. This indicates the excellent switchable superwettable characteristics of NH_3-C-P-U@Ti SSC with a separation efficiency of ~96.6%.

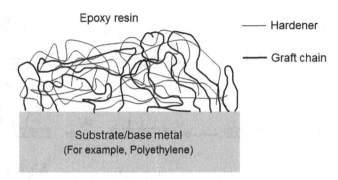

FIGURE 6.4 Representation of the surface interaction between grafted polymerized substrate and cure epoxy resin (thinner zigzag lines indicate hardener, while thicker black lines represent the grafted polymer chains).

Another important applying technique of SSCs for creating a switchable surface is grafting. Uyama et al. [17] reviewed the surface modification of polymer *via* grafting. The grafted layers of long polymer chains have controllable characteristics, high surface density, and precise localization of polymer chains at the surface (Figure 6.4). On this front, direct chemical modification, Ozone and γ-rays modification, glow discharge, UV-irradiation, and surface modification by electron beam have been extensively used. This surface functionalization by grafting was found to be effective in biomedical (for example, tissue adhesion), non-fouling applications, and slippery surfaces prevention. One of the important characteristics of the grafting technique is that while the surface properties are changed (i.e., switchability) upon the application of grafted polymer layers, the bulk properties of the substrate remain unchanged.

While the grafting technique effectively modifies the surface of polymers, the potential use of polymer brushes was found to be an effective media to create the new surfaces with multifunctionalities such as switchable and auto-response characteristics environmental triggers. This technique finds a unique niche in widespread applications such as resistance and response against lubrication and friction and the creation of bio-compatible surfaces with superior adhesion. These polymeric brushes (either homogeneous monomers or heterogeneous binary polymers) can be used in the form of ultra-thin polymeric layers, which further tune the physicochemical characteristics of the surface, such as wettability, adhesion chemical composition, and creation of switchable and auto-responsive surfaces. For instance, Uhlmann and co-workers [18] systematically explained the creation of ultra-thin switchable surfaces and their amplification of switching (for example, from superhydrophilic to superhydrophobic and superhydrophilic to ultrahydrophobic) *via* the application of binary polymer brushes. The synthesis of binary brushes comprises two-step grafting process, initiating "grafting from" followed by "grafting to", as shown schematically in Figure 6.5.

The grafting-from technique comprises initiation of polymerization at the surface of a substrate and initiating groups (normally azo-initiator) followed by

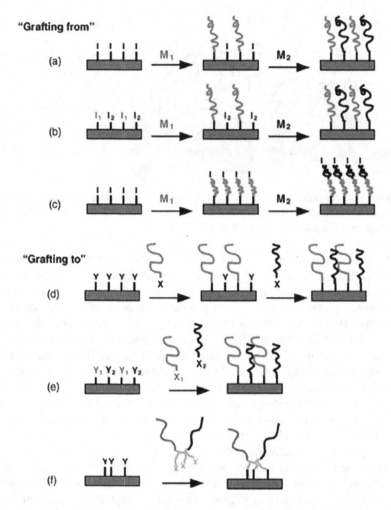

FIGURE 6.5 Schematic representation of the two-stage synthesis of binary polymer brush for the surface functionalization of polymers (a–c) grafting-from and (d–f) grafting-to approach. The process of grafting polymer brushes contains the following steps: (a) polymerization (two-step) at the surface with the same interior for both steps (slow initiation), (b) two-step polymerization at the interfaces with two different interiors, (c) controlled polymerization of di- or tri-block copolymer brushes creating a single interior with fast initiation (i.e., "grafting from"), (d) grafting of homo-polymer with same functional groups in two steps, (e) grafting of two different homo-polymer with different functional groups, and (f) termination of block co-polymers via a specific anchoring block. The abbreviation "I" represents initiator (azo-initiator), "M" specifies monomer (for example, styrene and 2-vinyl pyridine), "X" is the homo-polymer (for example, carboxy terminated poly(2-vinyl pyridine) or polystyrene), and "Y" denotes the molecules which form a layer of coupling functional groups (for example, 3-glycidoxypropyltrimethoxysilane). (Adopted and reprinted with permission from P. Ulhmann et al., *Progress in Organic Coatings*, 55 (2006) 168–174 [18].)

penetration of monomer molecules through the grafted polymer. Following this, the evolution of polymer brushes generally takes place at the interface. On the other hand, the grafting-to technique utilizes polymerization of end-functional polymer molecules, which further react with functional groups at the surface and form the tethered chains. While the grafting-from technique is advantageous in creating/tuning the switchable surface with high grafting density (~100 nm thick layer can be generated in single-go grafting), the grafting-to technique results in the creation of comparatively stable and homogeneous brushes due to the use of linear polymers with an additional advantage of narrow weight distribution. Both the techniques are useful for creating/tuning the surface characteristics (i.e., from superhydrophilic to ultrahydrophobic), shown in Figure 6.6 with some limitations. For example, the main limitation of the grafting-from technique is the generation of inhomogeneous surface, while low grafting densities limit the use of grafting-to techniques. Both the methods with some modifications can enhance the efficiency of switchable surfaces though their practical applications at commercial/bulk scale are still far from industrial acceptance. The related challenges regarding the above limitations are discussed subsequently in the below section.

A well-defined surface roughness could play a vital role in reaching the extreme amplification of hydrophilic and hydrophobic surfaces with switchable amplitude. This also intermingles the switchable and ultrahydrophobic characteristics of the surface. An ultrahydrophobic surface has minimum intrinsic contact angle (CA) and some surface roughness. The transition to ultrahydrophobicity occurs immediately at any surface roughness with minimum CA. Minko and coworkers [19] have implemented the idea of the ultrahydrophobic surface *via* creating a two-level structured surface with reversible switching from hydrophilic to ultrahydrophobic in response to external stimuli, as shown by the schematics in Figure 6.7. The first level on the PTFE surface was created *via* plasma etching by radio frequency oxygen plasma for ~600 s, which further reflected in needle-like

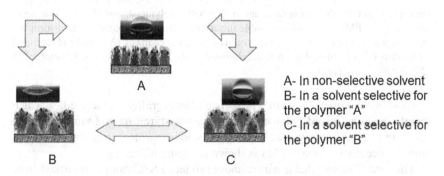

A- In non-selective solvent
B- In a solvent selective for the polymer "A"
C- In a solvent selective for the polymer "B"

FIGURE 6.6 Schematic representation of switchable physicochemical surface properties of a binary polymer brush: A – binary polymer brush made from hydrophilic polymer (for example, poly(2- vinylpyridine)) and B – hydrophobic polymer (e.g., polystyrene). (Adopted and reprinted with permission from P. Ulhmann et al., *Progress in Organic Coatings*, 55 (2006) 168–174 [18].)

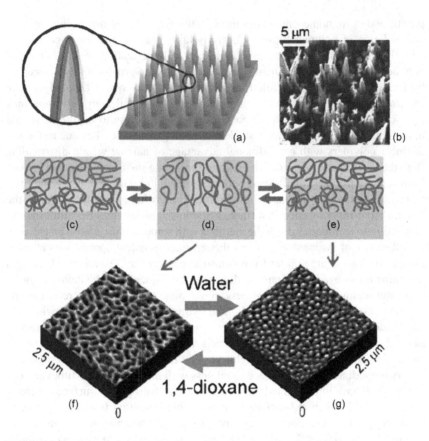

FIGURE 6.7 Schematic representation of the creation of switchable surface on polytetrafluoroethylene (PTFE): (a) needle-like surface morphology of switchable ultrahydrophobic surface, (b) SEM image of PTFE after plasma etching (for 600 s), (c–e) applying binary polymer brush (hydrophilic and hydrophobic polymers), and (f, g) atomic force microscope (AFM) images of switchable surface (i.e., from hydrophilic to ultrahydrophobic). The water droplets on the surface could be clearly seen in (g). (Adopted and reprinted with permission from S. Minko et al., *Journal of the American Chemical Society*, 125 (2003) 3896–3900 [19].)

morphology (Figure 6.7a and b). The second layer is grafted on the needles, which mainly contains mixed polymer brush (i.e., polystyrene/poly(2-vinylpyridine)), resulting in reversible switchable ultrahydrophobic surface, as characterized via atomic force microscope (AFM) as shown in Figure 6.7f and g.

The reversible switchable nature above produced SSC can be confirmed from the glancing amplitude from 20° in case of smooth surface to 150° for rough PTFE (Figure 6.8).

Interestingly, binary polymer brushes can also be used to create switchable adaptive surfaces. For instance, Ionov and co-workers [20] developed a reversible switchable surface pattern (i.e., adaptive and erasable, when the need arises)

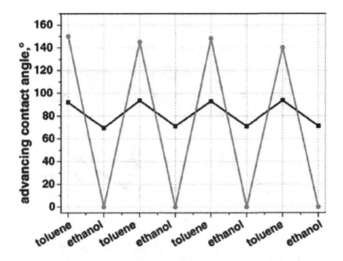

FIGURE 6.8 The CA of water for two different surfaces: (a) smooth silicon wafer (indicated by black colour with square (online version)) and (b) rough etched PTFE (indicated by orange colour with circles (online version)). (Adopted and reprinted with permission from P. Ulhmann et al., *Progress in Organic Coatings*, 55 (2006) 168–174 [18].)

via binary polymer brushes treated with solvent selective homo-polymer (photo-cross-linkable) and subsequently irradiated through photo-mask, as shown in Figure 6.9. This double layer coating has the unique advantage of a reversible switchable surface. The irradiated polymer chains of homo-polymer form the cross-linked pattern, while the polymer brushes at the non-irradiated region tend to switch the surface. Due to the non–cross-linking of polymers at the non-irradiated region, the polymer may swell and segregate. In contrast, cross-linked polymers cannot switch and hence maintain their original state. The repetitive treatment with selective solvent can observe the erasable patterns photo-cross-linked homo-polymer. The representative SEM images shown in Figure 6.10 indicate the adaptive reversible switchable surface created *via* irradiation of a polyisoprene/poly(2-vinylpyridine) brush through a fork (shaped) photo-mask. The patterns were developed *via* repetitive washing with dilute acid (pH ~2), followed by drying and water droplet deposition. The non-wetting of the irradiated region is obvious from Figure 6.10a, while water droplets are clearly observable on non-irradiated surfaces. Subsequent washing with ethanol erases the patterns, which is further evinced from low contrast (or no contrast) after the treatment with water vapour. This development and erasing of the film are reversible and can be repeated several times upon the external stimuli environment. The presence of switchable micro-channel was further confirmed from surface topography *via* AFM. As the distance of the grid from the surface increases (at a distance ~100 μm), an array of water droplets (size ~25 μm) is clearly visible, while the pattern is developed upon the overlapping of the grid with brush surface, thus indicating its reversible switchable characteristics (Figure 6.11).

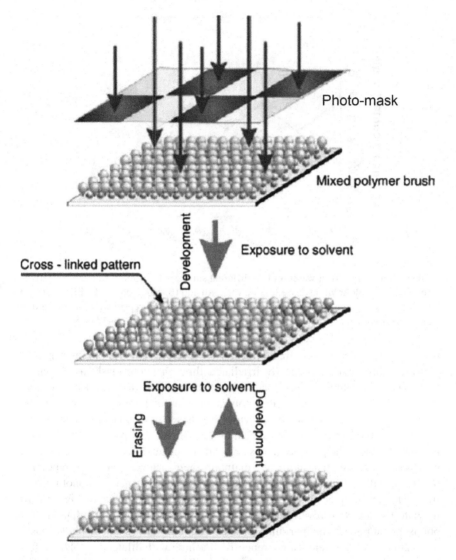

FIGURE 6.9 Schematic illustration of the creation of adaptive surface layer *via* photo-lithography of binary polymer brushes. (Adopted and reprinted with permission from L. Ionov et al., *Journal of the American Chemical Society*, 125 (2003) 8302–8306 [20].)

6.2.2 OVERVIEW OF APPLICATIONS TO THE MEDICAL SECTOR

Besides their applications to marine (non-fouling and oil/water separation) and transportation industries, SSCs also find a unique role in medical (or clinical) applications. Owing to their double-acting (or switchable) characteristics, these SSCs are the potential candidates for creating an antibacterial switchable surface

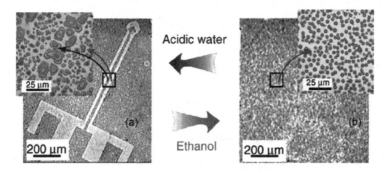

FIGURE 6.10 Representative SEM images of adsorption of water droplets on polymer brushes: (a) developed patterns and (b) erased patterns. (Adopted and reprinted with permission from L. Ionov et al., *Journal of the American Chemical Society*, 125 (2003) 8302–8306 [20].)

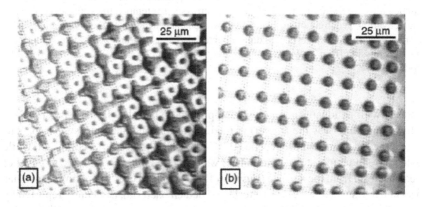

FIGURE 6.11 AFM topography of the surface of binary PI/P2VP brushes irradiated through a grid indicating reversible switchable characteristics: (a) grid in contact with brush and (b) grid at ~100 μm distance apart from the surface of brush. (Adopted and reprinted with permission from L. Ionov et al., *Journal of the American Chemical Society*, 125 (2003) 8302–8306 [20].)

(i.e., bacteria killing and removal). The SSCs used for the antibacterial purpose are of three types; (i) the coatings comprised of polymer derivatives that can switch/alter their chemical structures, (ii) coatings that are responsive to external stimuli, and (iii) coatings which comprises above two characteristics (switchable and responsive). The first work on the former type is reported by Chen et al. [21], wherein they have prepared polycarboxybetaine derivative containing ester group with dominant biocidal activity. The increase in pH of the surface (i.e., triggering) reflected in the removal of the ester group *via* hydrolysis process, thereby altering the properties of the surface (i.e., antifouling) to detach the killed bacteria (Figure 6.12a). While in the latter type, Yu et al. [22] grafted the silicon

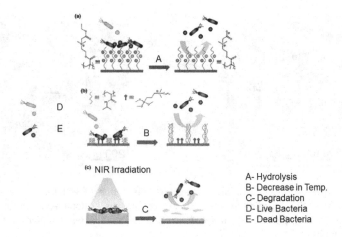

FIGURE 6.12 Schematic representation of development phases in SSCs for antibacterial applications: (a) switching the surface via hydrolysis, (b) temperature-controlled triggering/activation/switching of surface, and (c) switchable surface with the active and passive response having kill bacterial and detach bacterial characteristics. (Adopted and reprinted with permission from T. Wei et al., *Current Opinion in Chemical Engineering*, 1 (2021) 100727 [1].)

surface with nanopatterned poly(N-iso-propylacrylamide) (PNIPAAm) brushes. The temperature was the triggering medium to respond and auto-switch the surface. Once the temperature reached 37°C, the PNIPAAm chains collapsed, while a decrease in temperature (up to ~4°C) was reflected in the systematic removal of dead bacteria from the surface (Figure 6.12b). The third type of coating, which has a switchable surface (i.e., kill bacteria and detach bacteria), is recently reported by Qu and co-workers [23]. They have prepared the coatings by sequential deposition of gold nanoparticles along with photothermal effect followed by deposition of a layer of ascorbic acid phase-transitioned lysozyme (Figure 6.12c). Subsequently, near-infrared laser treatment on the surface could destroy and repel the attached bacteria.

Developing antibacterial SSCs based on different kill and detach bacteria mechanisms is a timely topic. Several studies that have been made in the past decade reported the development of diverse SSCs based on external stimuli for activating the switching characteristics of the surface, which typically includes triggering (activation or switching) *via* pH and temperature [24–26], the activity of chemical agents and their combination (i.e., salts, sugar, surfactants, and counterions) [27–31], and electric field [32].

Previously, the applications of antibacterial SSCs were limited to some standard substrates such as gold and silicon. Therefore, applying such coatings on medical devices or implants made of polymeric materials was a challenging task. However, in the past decade, notable progress has been made in developing some universal deposition/applying techniques with extended practical materials/substrates.

For instance, Wei et al. [33] have implemented layer by layer deposition technique to produce antibacterial multifunctional SSCs. The substrate material was first subjected to positive charge treatment followed by multilayer deposition of poly-anion, poly(acrylic acid-co-1-adamantan-1-ylmethylacrylate) film incorporated with apolycation, poly-(allylamine hydrochloride), and cationic b-cyclodextrin derivatives, which exhibits the biocidal activity. The removal of dead bacteria was performed subsequently by treating the surface with surfactant. This coating exhibited excellent switchable characteristics with extended applications on various substrates.

Interestingly, ring-opening reaction coupled with functional polymer epoxy groups, a simplest and robust deposition technique, was implemented by Mao et al. [34] for producing antibacterial SSCs on polymeric, metallic, and inorganic medical substrates. The deposited coating contains the first layer of copolymer quaternized glycidylmethacrylate (GMA) followed by another layer of 3-(dimethyl (4-vinylbenzyl)ammonium) propyl sulfonate (DVBAPS). The main function of GMA is to kill the bacteria, while DVBAPS dissociates the dead bacteria from the surface. Although the above methods serve the purpose of applying SSCs on different medical substrates, a surface pretreatment (i.e., induction of positive charge (or layer)) is indeed a requirement for enhancing the effectiveness of the deposited layer. In view of this, Wang and co-workers [35] used a tannic acid/Fe ion (TA/Fe) complex for developing the photothermal coatings with switching characteristics. The (TA/Fe) complex served the purpose of surface modification owing to its excellent adhesion, hydrophobic and electrostatic interaction, and metal coordination characteristics.

As mentioned earlier, several multifunctional SSCs have been developed with different deposition/applying techniques. However, the practical applications of these coatings are still a timely topic. In view of this, the current challenges and future perspectives associated with the synthesis and applying techniques of these SSCs are briefed below.

6.3 SMART SWITCHABLE COATINGS (SSCs): RELATED ISSUES, CHALLENGES, AND FUTURE PERSPECTIVES

While most of the research in the literature is focused on the applications of SSCs on metallic bioimplant surfaces, the practical applications of such coatings in polymeric devices and inorganic implant materials are yet to be strongly investigated. This is due to the fact that deposition of these coatings is a challenging task and requires sophisticated/precise instruments. Although few studies in the past reported the universal deposition techniques, these coatings are far from practical applications on a bulk scale. Furthermore, one of the important issues in the deposition of SSCs is the surface pretreatment by active agents, which can enhance their multifunctional performance *via* creating a positive ion-induced layer on the switchable surface. This eventually tunes the adhesion and dual-action switchability of the polymer surface. Another important issue related to the fabrication of SSCs is the reduction in deposition stages. Most of the reported SSCs thus far utilize the two-stage or multistage fabrication process, which is tedious and time consuming. Therefore, it would be prudent to develop such techniques that can prepare these SSCs in a single stage without compromising the multifunctionality and durability.

The profound discussion provided above allows us to suggest the use of indigenous stimuli in SSCs in future. These indigenous stimuli (particularly biological) may act as self-trigger in response to the external environmental changes (i.e., biological changes). The applications of SSCs reported so far are only limited to surface level, and no single study reports the eradication of bacteria underneath the surface. Therefore, it is reasonable to develop SSCs in future, which can kill and remove the bacteria inside a thin biofilm. The long-term storage and durability of these SSCs in an aggressive repeated (dynamic) environment is also an important problem that could be addressed in the future.

6.4   CONCLUSIONS

The ever-increasing demands for the antibacterial and antifouling surface in medical (particularly clinical translation) and marine sectors (e.g., oil/water separation during wastewater treatment) have led scientists to look for an alternative to conventional coatings. In view of this, recently, various coatings have been developed for medical and marine applications. However, SSCs are perhaps the most promising and potential candidates, keeping antibacterial and antifouling applications in view. By applying SSCs, the surface properties of substrates such as wettability, tribology, adhesion, lubrication, and most importantly, biocompatibility can be dramatically tuned. Therefore, such SSCs are gaining significant attention from the entire coatings community for surface functionalization. The discussion in the present chapter provides substantial insights about the synthesis and applying techniques of SSCs with their advantages and limitations.

The various applying techniques such as grafting-from, grafting-to, layer by layer deposition, polymer brush, porous materials, adaptive layering, and separation by filter paper or zeolite (the simplest technique with existing materials in nature) were found to be effective for creating a switchable (or reversible switchable in some cases) surfaces. One of the important parameters to consider while designing the SSCs for antibacterial applications is the dual-action characteristics of switchable surfaces (i.e., bacteria killing and removal), while wettability (hydrophobic/philic) of the surface is of great concern in marine applications.

In future, there is much more scope for the design and development of SSCs based on self-triggering mechanisms as most of the SSCs reported thus far utilize exogenous stimuli. The use of active agents and their incorporation is expected to enhance the functional performance of SSCs. Furthermore, no single study reports the applications of these coatings in an aggressive variable environment. The current trends in research make the SSCs sit squarely into the nexus of practical (or industrial) applications in the upcoming decade.

REFERENCES

1. T. Wei, Y. Qu, Y. Zou, Y. Zhang, Q. Yu, Exploration of smart antibacterial coatings for practical applications, *Current Opinion in Chemical Engineering*, 1 (2021) 100727.
2. K.G. Neoh, M. Li, E.T. Kang, E. Chiong, P.A. Tambayah, Surface modification strategies for combating catheter-related complications: Recent advances and challenges, *Journal of Materials Chemistry B*, 5 (2017) 2045–2067.
3. Z.K. Zander, M.L. Becker, Antimicrobial and antifouling strategies for polymeric medical devices, *ACS Macro Letters*, 7 (2017) 16–25.
4. B. Song, E. Zhang, X. Han, H Zhu, Y. Shi, Z. Cao, Engineering and application perspectives for designing an antimicrobial surfaces, *ACS Applied Materials and Interfaces*, 12 (2020) 21330–21341.
5. Q. Yu, Z. Wu, H. Chen, Dual-function antibacterial surface for biomedical applications, *Acta Biomaterialia*, 16 (2015) 1–13.
6. Y. Zou, Y. Zhang, Q. Yu, H. Chen, Dual function antibacterial surface to resist and kill the bacteria: painting the picture with two brushes simultaneously, *Journal of Materials Science and Technology*, 70 (2021) 24–38.
7. G. Cheng, H. Xite, Z. Zhang, S.F. Chen, S.Y. Jiang, A switchable biocompatible polymer surface with self-sterilizing and nonfouling capabilities, *Angewandte Chemie International Edition*, 47 (2008) 8831–8834.
8. X. Zhang, C. Wang, X. Liu, J. Wang, C. Zhang, Y. Wen, A durable and high-flux composite coating nylon membrane for oil-water separation, *Journal of Cleaner Production*, 193 (2018) 702–708.
9. L. Feng, Z. Zhang, Z. Mai, Y. Ma, B. Liu, L. Jiang, D. Zhu, A super-hydrophobic and super-oleophilic coating mesh film for the separation of oil and water, *Angewandte Chemie International Edition*, 43 (2004) 2012–2014.
10. Z. Xue, S. Wang, L. Lin, L. Chen, M. Liu, L. Feng, L. Jiang, A novel superhydrophilic and underwater superoleophobic hydrogel-coated mesh for oil/water separation, *Advanced Materials*, 23 (2011) 4270–4273.
11. Z. Xue, Y. Cao, N. Liu, L. Feng, L. Jiang, Special wettable materials for oil/water separation, *Journal of Materials Chemistry A*, 2 (2014) 2445–2460.
12. Z. Chu, Y. Feng, S. Seeger, Oil/water separation with selective superantiwetting/ superwetting surface materials, *Angewandte Chemie International Edition*, 54 (2015) 2328–2338.
13. R.K. Gupta, G.J. Dunderdale, M.W. England, A. Hozumi, Oil/water separation techniques: a review of recent progresses and future directions, *Journal of Materials Chemistry A*, 5 (2017) 16025–16058.
14. J. Yong, Y. Fang, F. Chen, J. Huo, Q. Yang, H. Bian, G. Du, X Huo, Femtosecond laser ablated durable superhydrophobic PTFE films with micro-through-holes for oil/water separation: separating oil from water and corrosive solutions, *Applied Surface Science*, 389 (2016) 1148–1155.
15. H. Bian, J. Yong, Q. Yang, X. Hou, F. Chen, Simple and low-cost oil/water separation based on the underwater superoleophobicity of the existing materials in our life or nature, *Frontiers in Chemistry*, 8 (2020) 1–7.
16. W. Liu, Y. He, Y. Zhang, Y. Liu, L. Long, F. Shen, G. Yang, X. Zhang, S. Zhang, S. Deng, A novel smart coating with ammonia-induced switchable superwettability for oily wastewater treatment, *Journal of Environmental Chemical Engineering*, 8 (2020) 104164.
17. Y. Uyama, O.K. Kat, Y. Ikada, Surface modification of polymer by grafting, *Advances in Polymer Science*, 137 (1998) 1–39.

18. P. Ulhmann, L. Ionov, N. Houbenov, M. Nitschke, K. Grundke, M. Motornov, S. Minko, M. Stamm, Surface functionalization by smart coatings: stimuli-responsive binary polymer brushes, *Progress in Organic Coatings*, 55 (2006) 168–174.

19. S. Minko, M. Muller, M. Motornov, M. Nitschke, K. Grundke, M. Stamm, Two-level structured self-adaptive surfaces with reversibly tunable properties, *Journal of American Chemical Society*, 125 (2003) 3896–3900.

20. L. Ionov, S. Minko, M. Stamn, J.F. Gohy, R. Jérome, A. Scholl, Reversible chemical patterning on stimuli-responsive polymer film: environment-responsive lithography, *Journal of American Chemical Society*, 125 (2003) 8302–8306.

21. G. Cheng, H. Xue, Z. Zhang, S. Chen, S. Jiang, A switchable biocompatible polymer surface with self-sterilizing and nonfouling capabilities, *Angewandte Chemie International Edition*, 47 (2007) 8831–8834.

22. Q. Yu, J. Cho, P. Shivapooja, L.K. Ista, G.P. López, Nanopatterned smart polymer surfaces for controlled attachment, killing and release of bacteria, *Applied Materials and Interfaces*, 5 (2013) 9295–9304.

23. Y. Qu, T. Wei, J. Zhao, S. Jiang, P. Yang, Q. Yu, H. Chen, Regenerable smart antibacterial surfaces: full removal of killed bacteria: via a sequential degradable layer, *Journal Materials Chemistry B*, 6 (2018) 3946–3955.

24. T. Wei, W. Zhan, Q. Yu, H. Chen, Smart biointerface with photo switched functions between bactericidal activity and bacteria-releasing ability, *ACS Applied Materials and Interfaces*, 9 (2017) 25767–25774.

25. Y. Xie, S. Chen, Y. Qian, W. Zhao, C. Zhao, Photo-responsive membrane surface: switching from bactericidal to bacteria-resistant property, *Materials Science and Engineering C*, 84 (2018) 52–59.

26. T. Manouras, E. Koufakis, E. Vasilaki, I. Peraki, M. Vamvakaki, Antimicrobial hybrid coatings combining enhanced biocidal activity under visible-light irradiation with stimuli-renewable properties, *ACS Applied Materials and Interfaces*, 13 (2021) 17183–17195.

27. J. Wu, D. Zhang, L. Zhang, B. Wu, S. Xiao, F. Chen, P. Fan, M. Zhong, J. Tan, Y. Chu, J. Yang, Long-term stability and salt-responsive behavior of polyzwitterionic brushes with cross-linked structure, *Progress in Organic Coatings*, 134 (2019) 153–161.

28. W. Zhan, Y. Qu, T. Wei, C. Hu, Y. Pan, Q. Yu, H. Chen, Sweet switch: sugar-responsive bioactive surfaces based on dynamic covalent bonding, *ACS Applied Materials and Interfaces*, 10 (2018) 10647–10655.

29. T. Wei, W. Zhan, L. Cao, C. Hu, Y. Qu, Q. Yu, H. Chen, Multifunctional and regenerable antibacterial surfaces fabricated using a universal strategy, *ACS Applied Materials and Interfaces*, 8 (2016) 30048–30057.

30. J. Wu, D. Zhang, Y. Wang, S. Mao, S. Xiao, F. Chen, P. Fan, M. Zhong, J. Tan, J. Yang, Electric assisted salt-responsive bacterial killing and release of polyzwitterionic brushes in low-concentration salt solution, *Langmuir*, 35 (2019) 8285–8293.

31. Y. Zhou, Y. Zheng, T. Wei, Y. Qu, Y. Wang, W. Zhan, Y. Zhang, G. Pan, D. Li, Q. Yu, H. Chen, Multi stimulus responsive bio interfaces with switchable bio adhesion and surface functions, *ACS Applied Materials and Interfaces*, 12 (2020) 5447–5455.

32. L. Děkanovsky, R. Elashnikov, M. Kubiková, B. Vokatá, V. Švorčik, O. Lyutakov, Dual-action flexible antimicrobial material: switchable self-cleaning, antifouling, and smart drug release, *Advanced Functional Materials*, 29 (2019). doi:10.1002/adfm.201901880.

33. T. Wei, W. Zhan, L. Cao, C. Hu, Y. Qu, Q. Yu, H. Chen, Multifunctional and regenerable antibacterial surface fabricated using a universal strategy, *ACS Applied Materials and Interfaces*, 8 (2016) 30048–30057.

34. S. Mao, D. Zhang, Y. Zhang, J. Yang, J. Zhang, A universal strategy for controllable functionalized polymer surfaces, *Advanced Functional Materials*, 30 (2020) 2004633.
35. Y. Wang, T. Wei, Y. Qu, Y. Zhou, Y. Zheng, C. Huang, Q. Yu, H. Chen, Smart, photothermally activated, antibacterial surfaces with thermally triggered bacteria-releasing properties, *ACS Applied Materials and Interfaces*, 12 (2020) 21283–21291.

7 Smart Coatings
Environmental Aspects and Future Perspectives

7.1 GENERAL DISCUSSION

The onsets of coating development/formation itself poses two questions from their applications, economics, environmental and future prospect point of view:

- Is it possible to form ecofriendly (green) and non-toxic smart coatings (SCs) commercially?
- If the answer to the above question is "Yes", is it possible to produce such SCs with multifunctional properties, particularly for both fixed and mobile structural materials in defence, space, security, and biomedical applications?

Although several studies mentioned in the previous chapters have successfully produced SCs for various applications and service environments, the development of multifunctional SCs is still far from complete. The recent developments in SCs revealed the crystal-clear relationship between corrosion protection and coatings with the aid of nanotechnology. Recently, the transformation from conventional chromate-based SCs to green (ecofriendly) SCs is notable and reflects the durable service lives of SCs. This was possible because the coatings community have identified the potential use of (i) natural minerals such as ZnO, TiO_2, SiO_2, halloysite $(Al_2Si_2O_5 \cdot nH_2O)$, and zeolite; (ii) nonmetal inorganic inhibitors/capsules such as clay structures, zeolites, and hydrotalcite; and (iii) ceramet-based inhibitors. For instance, Dias et al. [1] produced Ce-incorporated NaX-zeolite SCs on AA2024 for active corrosion protection. They found that the corrosion protection ability of Ce-enriched coatings is two to three times more than normal zeolite SCs. Besides this, the detailed microstructural investigations also revealed that the intermetallic inclusions appear to be a primary reason for the formation of Ce-rich sites, which consequently enhances the barrier protection ability of SCs. In another study by Abdullayev and Lvov [2], clay nanotubes are incorporated in SCs as a sealing agent to stop the leakage of benzotriazole. Interestingly, this inorganic encapsulation provided a breakthrough for the study of releasing/triggering mechanisms on SCs. The benign effect of nano inhibitors/capsules/nanoparticles in SCs is mostly attributed to their excellent compactness and uniformity (particle/grain size upto a few nm), thus more surface coverage is possible [3–10] with high surface energies [11].

DOI: 10.1201/9781003200635-7

In line with this, encapsulation of healing agents (including micro and nano) in SCs opens up wide opportunities to develop a variety of coatings with multifunctional properties for specific applications. For instance, the combination of carrier sensitive, sensing inhibitor, and healing/repair agent (capsules) would be a great effort to develop the SCs with different releasing mechanisms and rates, thereby paving the ways for multifunctional SCs. In view of this, future studies could also be directed to examine the releasing kinetics of carrier and sensing agents. Based on the discussions provided in the previous chapters, the authors also see tremendous scope in developing SCs based on the pH and this is one of the important stimuli/triggering mechanisms, as pH gradients are ultimately responsible for the redox electrochemical process.

The acknowledged need in SCs from marine structure, clinical translation, and bioimplant applications is the encapsulation of antifouling inhibitors/agents in healing/repairing smart carriers. These agents are expected to detect the corrosion early and mitigate it, thereby providing durable life. Furthermore, reducing the number of coats (compared to conventional coatings) and applying a thinner layer also brings a tremendous scope in developing SCs with versatile and flexible chemistry (which might help to reduce the number of coats/layers) for high throughput corrosion protection performance. For example, the use of siloxanes in SCs is expected to enhance the surface properties such as superhydrophobicity, self-cleaning, and ice-repellent characteristics [12]. Keeping this in view, there is a growing need to develop such SCs, particularly targeting water repellent applications, including hydrophobicity and superhydrophobicity.

The corrosion protection of automotive parts and aerospace materials is a hot and timely topic, which is of keen interest to researchers in the present century. Not only from the functional performance enhancement point of view but coatings should also serve the purpose of decoration in such structures/materials. These applications typically need coatings with smooth surface functionalities and a pleasant aesthetic look. On this front, waterborne soft-touch SCs are being researched on a priority basis. However, only a few have been accepted for the industrial and international standards. Besides superior corrosion protection, these SCs essentially provide enhanced resistance to abrasion and scratch to the materials/structures used in car/automobile interior and printable micro/nanoelectronic devices (i.e., MEMS and NEMS devices). The addition of cermet crystals such as ZnO, TiO_2 (rutile), and SnO_2 (cassiterite) and polymer inhibitors like fluoro-polymer are expected to provide the lustres look (from an aesthetic viewpoint), superhydrophobic, self-cleaning, and anti-dusting surface in architectural structures/materials [12]. However, developing such SCs with the above features is a challenging task, and these multifunctional aspects must meet the primary objective of corrosion protection.

Biodegradable organic SCs find a unique niche in bioimplants and bioresorbable materials due to their excellent compatible biointerference. At the same time, there is a growing need to develop/apply these SCs in a thin film form. Most of the biodegradable SCs contain polymer, which further enhances the multifunctionality with superior barrier protection, thus delaying the onsets of the corrosion process.

Therefore, it is anticipated to incorporate nanoparticles (may be metal oxide) in the near future for improved bone/cell adhesion, bone growth, longevity, and antibacterial activity. These SCs can potentially replace the conventional bioresorbable metallic coatings in bioimplants. Furthermore, advanced biodegradable organic SCs can also be used to control the drug delivery and for clinical translation applications while maintaining the synergistic combination of multifunctional properties and corrosion protection.

The purpose of this section is to get the readers familiar with recent developments in SCs for various applications and their future perspectives. In line with this, below we present some commercially available SCs with their important applications and producers in Table 7.1.

7.2 CHALLENGES IN THE DEVELOPMENT OF SCs

It's a great saying by anonymous that "Where there is a will, there is a way". We look at this proverb from the development point of view in a positive sense: "Where there is a development, there are challenges". We truly believe that the development in SCs is not exhaustive by any means. The main challenge in the field of SCs is their translation from research-based laboratory (micro/nano) scale to practical paint and coatings at industrial (meso) scale. This eventually requires knowledge about practical/industrial applications/requirements of SCs and their synthesis/fabrication techniques. A proper and collaborative dialogue between academia and industry is indeed a requirement to bridge the gap between technical/ scientific research and market demand.

Another key challenge from the life-cycle point of view of SCs is maintaining the synergistic combination of multifunctional performance and durable life. Often these SCs are subjected to the variable dynamic impact/loading conditions, which consequently deteriorate their functional and structural properties. In line with this, the need of producing chromate-free (or toxin-free inhibitor) SCs with enhanced toughness is the key priority of the coatings industry. However, the research is still at its development stage.

Maintaining multifunctional properties in SCs for a specific application is challenging. For instance, incorporating the carrier sensitives to different stimuli and different healing agents could open up novel ways to develop the SCs with healing and multifunctional properties [12]. Most of the SCs contain different healing agents based on a specific application. However, the studies related to the comprehensive understanding of healing kinetics, service life prediction of healing agents, and detection of early uncontrolled corrosion are still far from complete. In the current scenario, there is a continuous demand to develop a model for the predictive capabilities of healing agents, which can further identify the healing/repair capabilities of SCs to heal/repair the multiple damages.

For transportation industries, the utmost challenge is to develop the water repellent SCs with superhydrophobic and oleophobic surfaces. At the same time, these SCs should also be capable of multifunctionalities such as oil and ice repellant ability, minimum drag reaction, reduction in bio-fouling, and enhanced corrosion

TABLE 7.1

Summary of Commercially Available SCs for Glass, Architectural, Biomedical, Water Repellant, and Surface Cleaning/Antibacterial Applications

Products	Characteristic Features/ Multifunctionality	Applications	Producer/ Organization/ Company	Reference/Website
Intercept 8500 LPP	Antifouling, self-cleaning coating	Corrosion protection in marine structure/ vessels	BASF/AkzoNobel	[13]
PPG Paints	Self-cleaning, antibacterial, and anticorrosion	Corrosion protection in architectural structures	PPG	[14]
3M™ paint protection	Self-cleaning and anticorrosion, decorative	Hypershield in car decoration and anticorrosion	3M	[15]
Tyvek and Sorona	Antigraffiti, self-healing, and self-cleaning	Construction materials, graphics, and marine applications	Dow-DuPont	[16]
Deletum 5000 and 3000	Antigraffiti paints	Water and oil repellent applications	CG² Nano Coatings/Victor Castaño	[11,17]
SLIPS® foul protect, SLIPS® Dolphin, SLIPS® SeaClear	Self-cleaning, antibacterial, and antifouling	Water repellent applications and corrosion protection in the marine environment	Adaptive Surface Technologies	[18]
ECONTROL	Intelligent solar control glass	Glass shading applications	EControl-Glas GmbH & Co. KG	[11,19]
Ultra-Ever Dry	Superhydrophobic and oleophobic	Water repellent applications	UltraTech International, Inc.	[11,20]

protection performance. Furthermore, while designing the soft-touch surface coatings for aerospace and automobile interior applications, the main issue arises due to the manipulation of water-based coatings chemistries [12]. This eventually open up the insertion of required functional additives, which is indeed a difficult task. From the biomedical applications viewpoint, the development of advanced function bioresorbable SCs is an important open issue in the literature. The design of bioresorbable SCs opens up wide challenges regarding the incorporation of a variety of healing functionalities in SCs.

In summary, some of the major future challenges in the development of SCs that may affect the coating industries in general are as follows: (i) cost and greenhouse gases reduction, (ii) multifunctional properties in SCs, (iii) developments of SCs for fixed and moving infrastructure/materials, and (iv) multifunctional properties and corrosion protection synergy.

7.3   CONCLUSIONS

The sole purpose of this chapter is not to provide the mare summary on the previous chapters but to point out the conceivable future perspectives in the development of SCs with the aid of industrial applications. With ever increase in progress and recent developments in the field, undoubtedly SCs are the promising candidates for energy harvesting and pollution control devices for better sustainability. The main hurdle while maintaining the synergy between multifunctionality and corrosion protection performance in SCs is the increased production costs. The continuous developments are still in progress at the laboratory scale. However, their industrial acceptance is far from complete. A brief discussion is also provided on various challenges in the development of SCs mentioned in the previous chapters for various applications.

In summary, SCs provide a unique stimuli response and technological characteristics that are lagging in conventional coatings. From the profound reviews in the previous chapters, many approaches in the development of SCs with multifunctional characteristics at laboratory scale will find the industrial applications (at bulk scale) and their acceptance in the near future. To sum up, in single Layman's sentence: "SCs is the area of expectations that promises the new accomplishments in conventional coatings". With these common but important informative aspects (presented concisely), we conclude the book *Smart Coatings: Fundamentals, Developments and Applications*.

REFERENCES

1. S.A.S. Dias, S.V. Lamaka, C.A. Nogueira, T.C. Diamantino, M.G.S. Ferreira, Sol–gel coatings modified with zeolite fillers for active corrosion protection of AA2024, *Corrosion Science*, 62 (2012) 153–162.
2. E. Abdullayev, Y. Lvov, Clay nanotubes for corrosion inhibitor encapsulation: release control with end stoppers, *Journal of Materials Chemistry*, 20(32) (2010) 6681–6687.
3. G. Williams, H.N. McMurray, Inhibition of filiform corrosion on polymer coated AA2024-T3 by hydrotalcite-like pigments incorporating organic anions, *Electrochemical and Solid State Letters*, 7(5) (2004) B13–B15.
4. M. Kendig, M. Hon, A hydrotalcite-like pigment containing an organic anion corrosion inhibitor, *Electrochemical and Solid State Letters*, 8(3) (2005) B10–B11.
5. N.S. Bagal, V.S. Kathavate, P.P. Deshpande, Nano TiO_2 phosphate conversant coatings – a chemical approach, *Electrochemical Energy Technology*, De Gruyter, 4 (2018) 47–54.
6. V.S. Kathavate, N.S. Bagal, P.P. Deshpande, Corrosion protection performance of nano TiO_2 incorporated phosphate coatings obtained by anodic electrochemical treatment, *Corrosion Reviews*, 37(6) (2019) 565–578.

7. S.P.V. Mahajanarn, R.G. Buchheit, Characterization of inhibitor release from Zn-Al-V10O28 (6-) hydrotalcite pigments and corrosion protection from hydrotalcite-pigmented epoxy coatings, *Corrosion*, 64(3) (2008) 230–240.

8. V.S. Kathavate, D.N. Pawar, N.S. Bagal, P.P. Deshpande, Role of nano ZnO particles in the electrodeposition and growth mechanism of phosphate coatings for enhancing the anti-corrosive performance of low carbon steel in 3.5% NaCl aqueous solution, *Journal of Alloys and Compounds*, 823 (2020) 153812.

9. H.N. McMurray, G. Williams, Inhibition of filiform corrosion on organic coated aluminum alloy by hydrotalcite-like anion-exchange pigments, *Corrosion*, 60(3) (2004) 219–228.

10. V.S. Kathavate, P.P. Deshpande, Role of nano TiO_2 and nano ZnO particles on enhancing the electrochemical and mechanical properties of electrochemically deposited phosphate coatings, *Surface and Coatings Technology*, 394 (2020) 125902.

11. S.B. Ulaeto, J.K. Pancrecious, T.P.D. Rajan, B.C. Pai, Smart coatings, In: *Noble Metal-Metal Oxide Hybrid Nanoparticles*, pp. 341–372. Elsevier (2019). https://doi.org/10.1016/B978-0-12-814134-2.00017-6.

12. M.F. Montemor, Functional and smart coatings for corrosion protection: a review of recent advances, *Surface and Coatings Technology*, 258 (2014) 17–37.

13. https://www.basf.com/global/en/who-we-are/organization/locations/europe/german-sites/Muenster/News-Releases/BASF-completes-sale-of-its-industrial-coatings-businesses-to-AkzoNobel.html.

14. https://www.ppg.com/.

15. https://www.3m.com/3M/en_US/post-factory-installation-us/paint-protection-film/.

16. http://www.dow-dupont.com/.

17. https://www.cg2nanocoatings.com/victorcastano.shtml.

18. https://adaptivesurface.tech/marine/.

19. https://www.linkedin.com/redir/redirect?url=http%3A%2F%2Fwww%2Eecontrol-glas%2Ede&urlhash=q0zo&trk=about_website.

20. https://spillcontainment.com/.

Index

A

acrylic 1
active targeting 56
anion 18
anodic protection 3
antifouling and antimicrobial SCs 6, 64, 73,
    76, 82
antimicrobial 63
antireflectivity 45
autografts 48
automotive industries 1
azo-initiator 67

B

barrier protection 22
bioactivity 45, 49, 50
biocidal 63
biocompatibility 45, 49, 50, 58, 59
biodegradability or biodegradable 45,
    82, 83
biofouling 45, 63
bioglass 47
bioimplant materials 45
bioimplant stents 46
bioinspired materials 45
bioinspired smart coatings 45, 46
bone regeneration 48

C

capsulation 26
cathodic disbonding 5
cation/cationic 18, 35
charge transfer 19
chemical stability 4, 34
chemical stimuli response 8
chemical triggering 8
chromate conversion coatings 24
clinical translations 55, 82
coatings science 58
conducting polyaniline 2
contact angle 69
corrosion kinetics 18
corrosion sensing 34, 36

D

defect sensing SCs 6
dendrimers 26
Diels-Alder reaction 26
drug delivery 2, 55, 58, 59
Duco Finished 1
DuPont 1

E

electrochemical corrosion 6
electrochemical properties 31
electrodeposited coatings (e-coat) 1
electrode potential 19
electrophoretic deposition 23, 47
encapsulation 26
environmental aspect 81
environmental sensitive coatings 6
equilibrium potential 18
exchange current density 19
extrinsic defects 5

F

free energy 17
free radical 35

G

grafting from 67, 68
grafting to 67, 68
green inhibitors 7
glass transition temperature 4
glow discharge 67

H

hazardous air pollution 35
healing kinetics 83
hydrophobicity 82

I

impact loading 33
inhibition protection 22

87

Printed in the United States
by Baker & Taylor Publisher Services